Ansible快速入门
技术原理与实战

史晶晶 编著

电子工业出版社
Publishing House of Electronics Industry
北京·BEIJING

内 容 简 介

本书以新的自动化运维工具 Ansible 为主要内容，侧重于实战，由浅入深地介绍了 Ansible 以及周边产品 Ansible Galaxy 和 Ansible Tower 的用法。本书共计 6 章，前 4 章由浅及深、层层递进地介绍了 Ansible 的使用方法。第 5 章着重介绍了 Ansible 的代码分享机制 role 以及其分享平台 Ansible Galaxy。第 6 章概括性地介绍了企业级软件 Ansible Tower。全书的叙述风格通俗易懂，没有过多地引入复杂的概念，侧重于讲解原理，立足于实战，引领 Ansible 新手轻松入门。

未经许可，不得以任何方式复制或抄袭本书之部分或全部内容。
版权所有，侵权必究。

图书在版编目（CIP）数据

Ansible 快速入门：技术原理与实战 / 史晶晶编著. —北京：电子工业出版社，2017.6
ISBN 978-7-121-31502-2

Ⅰ. ①A… Ⅱ. ①史… Ⅲ. ①程序开发工具 Ⅳ. ①TP311.561

中国版本图书馆 CIP 数据核字（2017）第 105131 号

责任编辑：安　娜
印　　刷：北京虎彩文化传播有限公司
装　　订：北京虎彩文化传播有限公司
出版发行：电子工业出版社
　　　　　北京市海淀区万寿路 173 信箱　　邮编：100036
开　　本：787×980　1/16　　印张：11　　字数：192 千字
版　　次：2017 年 6 月第 1 版
印　　次：2023 年 7 月第 14 次印刷
定　　价：49.00 元

凡所购买电子工业出版社图书有缺损问题，请向购买书店调换。若书店售缺，请与本社发行部联系，联系及邮购电话：(010) 88254888，88258888。
质量投诉请发邮件至 zlts@phei.com.cn，盗版侵权举报请发邮件至 dbqq@phei.com.cn。
本书咨询联系方式：(010) 51260888-819，faq@phei.com.cn。

内容介绍

Ansible 是对机群进行软件安装、配置和应用部署的自动化工具。自 2012 年 Ansible 出现后，以其使用简单、功能实用等特点得到了广泛关注，成为自动化运维工具中的冉冉新星。仅三年后就被红帽（Red Hat）公司收购，目前受到众多软件公司的关注、推广和使用。自动化部署无论对系统管理员还是对软件开发人员来说，都会减少重复的手工操作，提高部署效率。Ansible 作为学习成本小、学习路径短的一款工具，更是值得了解和使用。

作为一本 Ansible 入门书籍，读者只需对 Linux 有最基本的了解就可以轻松读懂本书。

本书的内容共计 6 章，分为以下三个部分：

第一部分　Ansible 基本工具的讲解（第 1 章～第 4 章）

本书首先介绍了 Ansible 架构，然后讲解了 Ansible 的主要概念，接着又介绍了一些具体使用方法，步步递进、层层深入地介绍了 Ansible 的相关知识。

第 4 章对前面 3 章中的知识细节进行了补充，以便读者能够深入理解 Ansible 的基本使用方法。

第二部分　role 和衍生工具 Ansible Galaxy 的介绍（第 5 章）

role 是 Ansible 最为推荐的重用代码的方式，并为其开发了 Ansible Galaxy 代码分享网站。但因其概念较为复杂，所以对其单独进行讲解。

第三部分　企业级收费软件 Ansible Tower 介绍（第 6 章）

企业级用户面临着更加复杂的应用环境和更高的安全要求，Ansible Tower 就是一款解决企业级用户难题的收费软件。

代码的运行环境

建议读者安装 Linux 虚拟机作为 Ansible 的管理节点来测试本书中的代码。Ansible 目前已被红帽公司收购，对 Red Hat Linux 系统的支持较为完善，建议读者最好选择 Red Hat Linux 7 或 CentOS 7。

排版约定

为了使读者能快速把握到代码的重点，本书中代码的重要部分用加粗字体表示强调。

例如，下面的代码强调变量"ansible_os_family"的用法：

```
---
- hosts: all
  user: root
  tasks:
  - name: echo system
    shell: echo {{ ansible_os_family }}
  - name install ntp on Debian linux
    apt: name=git state=installed
    when: ansible_os_family == "Debian"
  - name install ntp on redhat linux
    yum: name=git state=present
```

勘误信息

笔者对本书中所有的代码都进行了完整的测试，书中的文字也经过了反复的斟酌。尽管如此，由于时间紧迫且作者水平有限，错误和疏漏难以避免，还需要广大读者的反馈和修订，以使得本书更加完善。因此，如果您发现书中的任何错误，小到错别字，大到代码运行错误，都希望您能及时反馈。您的任何一次勘误，都会令笔者和其他读者受益，再次表示感谢。

勘误地址：

https://github.com/ansible-book/errata

或者

http://getansible.com/reference/errata

反馈勘误方法：

提问题到 Github 项目上：https://github.com/ansible-book/errata

发邮件给笔者：shijingjing02@163.com

致谢

感谢同组（Labs 和 Insights）的同事对本书早期版本的肯定，使我受到鼓舞，有信心编辑成册。感谢我的经理赵东在公司内部推广宣传，感谢周兆林（Jaylin）对早期代码仔细认真的测试，感谢傅炜（网名：Tekkaman Ninja）多次对文字提出大量的修改建议。

这是笔者第一次写书，在写书的过程中耗费了大量的时间和精力，若是没有家人的鼎力支持，这本书根本无法完成。尤其感谢老公为我分担了怀孕和育儿的大量调研工作，使我能够有时间投入本书的写作中。

本书的后期写作和修订正值我怀孕分娩前后，时间上难免拖拉，感谢本书的编辑安娜对我的理解和支持。

轻松注册成为博文视点社区用户（www.broadview.com.cn），扫码直达本书页面。

- **提交勘误**：您对书中内容的修改意见可在 提交勘误 处提交，若被采纳，将获赠博文视点社区积分（在您购买电子书时，积分可用来抵扣相应金额）。
- **交流互动**：在页面下方 读者评论 处留下您的疑问或观点，与我们和其他读者一同学习交流。

页面入口：*http://www.broadview.com.cn/31502*

目录

第 1 章 Ansible 介绍 1

1.1　Ansible 介绍 2
1.2　Ansible 解决了什么运维痛点 2
1.3　架构 2
　　1.3.1　Ansible 的架构 2
　　1.3.2　Ansible Tower 的架构 4

第 2 章 Ansible 入门 6

2.1　安装 Ansible 7
　　2.1.1　在管理员的电脑上安装 7
　　2.1.2　被管理的远程主机 7
2.2　Ansible 管理哪些主机 8
　　2.2.1　什么是主机目录 8
　　2.2.2　主机目录配置文件 8
2.3　Ansible 用命令管理主机 9
　　2.3.1　Ansible 命令的格式 9
　　2.3.2　Ansible 命令的功能 9
2.4　Ansible 用脚本管理主机 10
　　2.4.1　执行脚本 Playbook 的方法 11
　　2.4.2　Playbook 的例子 11
　　2.4.3　Play 和 Playbook 13

目录 | VII

2.5 Ansible 模块 14
    2.5.1 什么是 Ansible 模块 14
    2.5.2 在命令行里使用模块 15
    2.5.3 在 Playbook 脚本中使用模块 15
    2.5.4 Ansible 模块的特点 15
    2.5.5 常用模块 16

第 3 章 Ansible 进阶 30

3.1 Ansible 的配置 31
    3.1.1 可以配置什么 31
    3.1.2 Ansible 配置文件的优先级 31

3.2 主机清单 32
    3.2.1 远程主机的分组 33
    3.2.2 设置连接参数 34
    3.2.3 变量 34

3.3 Ansible 的脚本 Playbook 36
    3.3.1 Playbook 的文件格式 YAML 36
    3.3.2 执行 Playbook 的命令 37
    3.3.3 Playbook 的基本语法 38
    3.3.4 变量 45
    3.3.5 Playbook 也有逻辑控制语句 53
    3.3.6 重用 Playbook 58
    3.3.7 用标签，实现执行 Playbook 中的部分任务 66

3.4 更多的 Ansible 模块 69
    3.4.1 模块的分类 69
    3.4.2 Extra 模块的使用方法 70
    3.4.3 命令行查看模块的用法 71

3.5 最佳使用方法 71
    3.5.1 写 Playbook 的原则 71
    3.5.2 参考别人的 Playbook 72

第 4 章 Ansible Playbook 杂谈 73

4.1 再谈 Ansible 变量 74
4.1.1 变量的作用域 74
4.1.2 变量的优先级 74

4.2 使用 lookup 访问外部文件或数据库中的数据 80
4.2.1 lookup 读取文件 81
4.2.2 lookup 生成随机密码 81
4.2.3 lookup 读取环境变量 82
4.2.4 lookup 读取 Linux 命令的执行结果 83
4.2.5 lookup 读取 template 变量替换后的文件 83
4.2.6 lookup 读取配置文件 84
4.2.7 lookup 读取 CSV 文件的指定单元 86
4.2.8 lookup 读取 DNS 解析的值 87
4.2.9 更多的 lookup 功能 91

4.3 过滤器 91
4.3.1 过滤器使用的位置 91
4.3.2 过滤器对普通变量的操作 92
4.3.3 过滤器对文件路径的操作 96
4.3.4 过滤器对字符串变量的操作 99
4.3.5 过滤器对 JSON 的操作 106
4.3.6 过滤器对数据结构的操作 109
4.3.7 过滤器的链式/连续使用 111

4.4 测试变量或表达式是否符合条件 111
4.4.1 测试字符串 112
4.4.2 比较版本 113
4.4.3 测试 List 的包含关系 113
4.4.4 测试文件路径 114
4.4.5 测试任务的执行结果 115

4.5 认识插件 117
4.5.1 插件类型 118
4.5.2 常用的插件介绍 119

第 5 章 role 和 Ansible Galaxy 123
5.1 role 和 Ansible Galaxy 的简要介绍 124
5.1.1 role 124
5.1.2 Ansible Galaxy 124
5.2 role 的放置位置 124
5.2.1 当前目录的 roles 文件夹下 124
5.2.2 环境变量 ANSIBLE_ROLES_PATH 定义的文件夹 125
5.2.3 Ansible 配置文件中 roles_path 定义的文件夹 125
5.2.4 默认文件夹/etc/ansible/roles 125
5.3 在 Playbook 中如何调用 role 126
5.3.1 调用最简单的 role 126
5.3.2 通过 pre_tasks 和 post_tasks 调整 role 和任务的顺序 127
5.3.3 调用带有参数的 role 129
5.3.4 与 when 一起使用 role 129
5.4 如何写 role 130
5.4.1 role 的完整定义 130
5.4.2 默认变量和普通变量的区别 131
5.4.3 tasks/main.yml 如何使用变量、静态文件和模板 132
5.5 role 的依赖 134
5.6 Ansible Galaxy 网站介绍 136
5.6.1 从 Ansible Galaxy 网站上下载 role 136
5.6.2 分享你的 role 139
5.7 演示 role 的创建和分享 139
5.7.1 改造单个的 Playbook 为 role 141
5.7.2 在 Ansible Galaxy 中分享 role 144

第 6 章 Ansible Tower 145

6.1 为什么要用 Ansible Tower 146
6.1.1 Ansible 和 Tower 的用户视角架构图 146
6.1.2 Ansible Tower 的主要功能 147

6.2 如何使用 Ansible Tower 149
6.2.1 安装方法 149
6.2.2 使用方法 152
6.2.3 总结 161

6.3 与第三方平台的整合 163
6.3.1 Ansible Tower API 163
6.3.2 Ansible Tower 提供的命令行工具 164

附录 A 166

第 1 章

Ansible 介绍

本章重点

1.1　Ansible 介绍
1.2　Ansible 解决了什么运维痛点
1.3　架构

1.1 Ansible 介绍

Ansible 是一个部署一群远程主机的工具。这里"远程主机（Remote Host）"是指任何可以通过 SSH 登录的主机，所以它既可以是远程虚拟机或物理机，也可以是本地主机。

Ansible 通过 SSH 协议实现管理节点与远程节点之间的通信。理论上来说，只要能通过 SSH 登录到远程主机来完成的操作，都可以通过 Ansible 实现批量自动化操作。

包括：复制文件、安装包、发起服务，等等。

1.2 Ansible 解决了什么运维痛点

Ansible 解决了如何大批量、自动化地实现系统配置、应用部署、命令和服务操作的问题。其脚本具有灵活、可重入的特性，极大地减少了运维人员的重复劳动，提高了运维效率。

1.3 架构

1.3.1 Ansible 的架构

Ansible 管理节点和远程主机节点间通过 SSH 协议进行通信。所以配置 Ansible 的时候，只需保证从 Ansible 管理节点通过 SSH 协议能够连接到被管理的远程节点即可。注意，SSH 必须配置为公钥认证登录方式，而非密码认证，第 2 章会讲到具体的配置方法。

1. 连接方式（SSH）

在管理节点安装 Ansible 及所依赖的软件。由于管理节点需通过 SSH 连接被管理的主机来执行命令或者脚本，因此被管理的远程节点需要配置并启用 SSH 服务，此外无须安装其他特殊软件。管理节点只在执行脚本或命令时与远程主机连接，没有特别的同步机制，所以发生断电等异常时一般不会影响 Ansible。

2. 支持多种类型的主机

Ansible 可以同时管理 Red Hat 系的 Linux、Debian 系的 Linux 以及 Windows 主机。Ansible 的工作原理如图 1.1 所示。

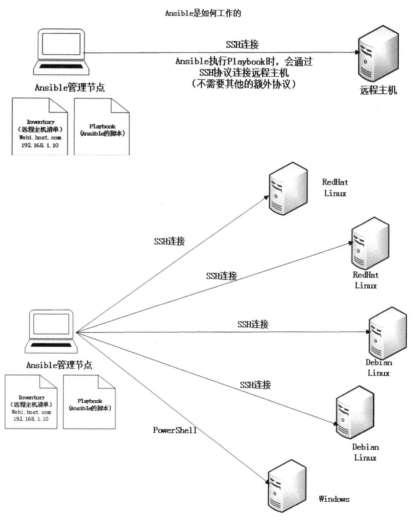

图 1.1　Ansible 工作原理

1.3.2 Ansible Tower 的架构

1. Ansible Tower 的由来

Ansilbe Tower 是一款针对企业用户的收费软件。

在 1.3.1 节的 Ansible 架构和第 2 章 Ansbile 的安装中会讲到,每一台被 Ansible 远程管理的主机,都需要配置采用公钥认证的 SSH 连接。密钥的配置和维护对于管理几台虚拟机和远程主机的个人用户不会有什么问题,但对于拥有大量主机和管理人员的企业用户来说,则可能有以下困扰。

- 维护工作量大:每增加一台主机,都需要手工配置一下 SSH 连接,企业级的 PC 主机成百上千,每个管理员都需要在自己的电脑上配置所有的 SSH 连接,无疑工作量巨大。
- 安全隐患大:在安全方面如果管理员能够拿到 SSH 私钥,或者复制给别人,那么对于生产环境来说,无疑是最大的安全隐患。
- 协同性弱:每个管理员都可能有自己的脚本库,有的脚本可能是为了解决同一个问题,但管理员之间没有一个通用的协作共享脚本的机制。
- 可视性差:基本一般通过 Shell 终端运行,对于大量主机批量配置的情况,其执行状态的表现能力有限,不利于后期对于结果的统计和分析。

Ansilbe Tower 是一款针对企业用户的收费软件,功能强大,很好地解决了以上困扰。

2. Ansible Tower 能做什么

Ansible Tower 则是针对企业用户环境、中心化的 Ansible 管理节点,它向管理员提供网页接口,来运行 Ansible 脚本 Playbook。

- 管理员在 Ansible Tower 上使用和分享主机的 SSH 私钥,但是不能查看和复制私钥文件。
- Ansible 网站上的所有管理员都可以共享 Playbook 脚本,减少重复工作。
- 此外,Ansible Tower 还可以收集和展现所有主机的 Playbook 的执行状况,以便于统计和分析主机的状态。

Ansible Tower 架构图如图 1.2 所示。

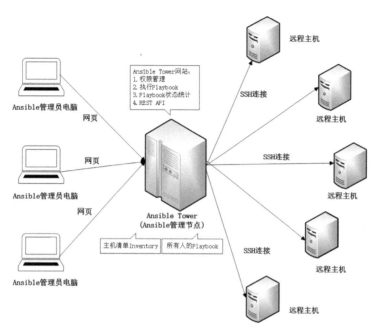

图 1.2 Ansible Tower 架构图

第 2 章

Ansible 入门

本章重点

2.1 安装 Ansible
2.2 Ansible 管理哪些主机
2.3 Ansible 用命令管理主机
2.4 Ansible 用脚本管理主机
2.5 Ansible 模块

2.1 安装 Ansible

这里以 Red Hat 系 Linux 为例，如果是使用 Windows 或者 Mac 的读者，也建议你在 Red Hat 系的虚拟机下安装并体验 Ansible。其他系统的安装方法请参考 Ansible 官网。

2.1.1 在管理员的电脑上安装

1. 安装 Ansible 软件

```
$ # Redhat/CentOS Linux 上，Ansible 目前放在的 epel 源中
$ # Fedora 默认源中包含 Ansible，直接安装包即可
$ sudo yum install epel-release
$ sudo yum install ansible -y
```

2. 配置 Ansible 管理节点和主机的连接

其实就是配置从**管理节点到远程主机**之间基于密钥（无密码的方式）的 SSH 连接。

```
$ # 生成 SSH 密钥
$ ssh-keygen
$ # 复制 SSH 密钥到远程主机，这样 SSH 的时候就不需要输入密码了
$ ssh-copy-id remoteuser@remoteserver
$ # SSH 的时候不会提示是否保存密钥
$ ssh-keyscan remote_servers >> ~/.ssh/known_hosts
```

验证 SSH 配置：在管理节点执行下面的 SSH 命令，既不需要输入密码，也不会提醒你存储密钥，那就成功啦。

```
$ ssh remoteuser@remoteserver
```

2.1.2 被管理的远程主机

不需要安装特殊的包，只需配置并启动 SSH 服务且 Python 版本在 2.4 以上即可，Red Hat Linux 一般安装方式都是默认安装的。

2.2 Ansible 管理哪些主机

Ansible 需要知道自己可以管理的主机有哪些，以及一些连接参数，它由主机目录配置文件来管理。

2.2.1 什么是主机目录

主机目录（Host Inventory，又称主机清单）是配置文件，用来告诉 Ansible 需要管理哪些主机，并且把这些主机按需分类。

例如，可以根据用途分类为数据库节点、服务节点等，也可以根据地点分类为中部机房、西部机房等。

2.2.2 主机目录配置文件

默认的文件是：/etc/ansible/hosts。

可以修改为其他文件，第 3 章 Ansible 进阶中将介绍使用其他文件路径作为主机目录文件的方法。

最简单的 Hosts 文件示例。

```
192.168.1.50
aserver.example.org
bserver.example.org
```

带分组的 Hosts 文件示例。

```
[webservers]
foo.example.com
bar.example.com

[dbservers]
one.example.com
two.example.com
three.example.com
```

2.3 Ansible 用命令管理主机

Ansible 提供了一个命令行工具，在官方文档中给命令行起了一个名字叫 Ad-Hoc Commands。

2.3.1 Ansible 命令的格式

Ansible 命令的格式是：

```
ansible <host-pattern> [options]
```

2.3.2 Ansible 命令的功能

先不用深究 Ansible 命令的语法，介绍完模块（2.5 节）后，就可以理解语法了。先从感观上，通过下面的命令体会一下 Ansible 命令行的功能。

1. 检查 Ansible 安装环境

检查所有的远程主机，是否以"bruce"用户创建了 Ansible 管理主机可以访问的环境：

```
$ansible all -m ping -u bruce
```

2. 执行命令

在所有的远程主机上，默认以当前 Bash 的同名用户，在远程主机执行"echo hello"：

```
$ansible all -a "/bin/echo hello"
```

3. 复制文件

复制文件 /etc/host 到远程主机（组）"Web"，位置为 /tmp/hosts：

```
$ ansible web -m copy -a "src=/etc/hosts dest=/tmp/hosts"
```

4. 安装包

在远程主机（组）WebServers 上安装 yum 包：

```
$ ansible web -m yum -a "name=acme state=present"
```

5. 添加用户

```
$ ansible all -m user -a "name=foo password=<crypted password here>"
```

6. 下载 Git 仓库

```
$ ansible web -m git -a "repo=git://foo.example.org/repo.git dest=/srv/myapp version=HEAD"
```

7. 启动服务

```
$ ansible web -m service -a "name=httpd state=started"
```

8. 并行执行

启动 10 个并行执行重启。

```
$ansible lb -a "/sbin/reboot" -f 10
```

9. 查看远程主机的全部系统信息

```
$ ansible all -m setup
```

2.4　Ansible 用脚本管理主机

为了避免重复地输入命令，Ansible 提供了脚本功能。Ansible 脚本的名字叫 Playbook，使用的是 YAML 格式，文件以 yml 或 yaml 为后缀。

注：YAML 和 JSON 类似，是一种表示数据的文本格式，其含义是"YAML Ain't a Markup Language"，即 YAML 不是一种标记语言的递归缩写。

2.4.1 执行脚本 Playbook 的方法

```
$ansible-palybook deploy.yml
```

2.4.2 Playbook 的例子

Playbook 包含了几个关键字,每个关键字的含义如下。

- **hosts**:某主机的 IP,或者主机组名,或者关键字 all。
- **remote_user**:以某个用户身份执行。
- **vars**:变量。
- **tasks**:Playbook 的核心,定义顺序执行的动作 Action。每个 Action 调用一个 Ansible 模块。

❶ action 语法:

```
module: module_parameter=module_value
```

❷ 常用的模块有 yum、copy、template 等,模块在 Ansible 中的作用,相当于 bash 脚本中的 yum、copy 这样的命令,2.5 节会介绍。

- **handers**:Playbook 的 Event 处理操作,有且仅有在被 Action 触发时才会执行。但多次触发只执行一次,并按照声明的顺序执行。

以下是一个为主机(组)Web 部署 Apache 的 deploy.yml 文件,部署步骤如下:

- 安装 Apache 包。
- 复制配置文件 httpd,并保证复制文件后,Apache 服务会被重启。
- 复制默认的网页文件 index.html。
- 启动 Apache 服务。

```
---
- hosts: web
  vars:
    http_port: 80
    max_clients: 200
  remote_user: root
  tasks:
  - name: ensure apache is at the latest version
```

```yaml
    yum: pkg=httpd state=latest

  - name: Write the configuration file
    template: src=templates/httpd.conf.j2 dest=/etc/httpd/conf/httpd.conf
    notify:
    - restart apache

  - name: Write the default index.html file
    template: src=templates/index.html.j2 dest=/var/www/html/index.html

  - name: ensure apache is running
    service: name=httpd state=started
  handlers:
    - name: restart apache
      service: name=httpd state=restarted
```

如果读者对YAML不熟悉，但有一定的JSON基础，则把上面的deploy.yml格式转化为JSON格式后如下所示。

```
[
 {
   "hosts": "web",
   "vars": {
     "http_port": 80,
     "max_clients": 200
   },
   "remote_user": "root",
   "tasks": [
     {
       "name": "ensure apache is at the latest version",
       "yum": "pkg=httpd state=latest"
     },
     {
       "name": "Write the configuration file",
       "template": "src=templates/httpd.conf.j2 dest=/etc/httpd/conf/httpd.conf",
       "notify": [
         "restart apache"
```

```
      ]
    },
    {
      "name": "Write the default index.html file",
      "template": "src=templates/index.html.j2 dest=/var/www/html/index.html"
    },
    {
      "name": "ensure apache is running",
      "service": "name=httpd state=started"
    }
  ],
  "handlers": [
    {
      "name": "restart apache",
      "service": "name=httpd state=restarted"
    }
  ]
}
]
```

提供 JSON 和 YAML 互转的在线网站为：http://www.json2yaml.com/。

2.4.3 Play 和 Playbook

Playbook 是指一个可被 Ansible 执行的 YAML 文件，其结构如下所示。

```
---
- hosts: web
  remote_user: root
  tasks:
  - name: ensure apache is at the latest version
    yum: pkg=httpd state=latest
```

其实在一个 Playbook 文件中，还可以针对两组主机进行不同的操作，例如，给 web（组）安装 HTTP 服务器，以及给 lb（组）安装 MySQL 并放在一个文件中。

```
---
#安装 apache 的 Play
```

```
- hosts: web
  remote_user: root
  tasks:
  - name: ensure apache is at the latest version
    yum: pkg=httpd state=latest

# 安装 MySQL Server 的 Play
- hosts: lb
  remote_user: root
  tasks:
  - name: ensure mysqld is at the latest version
    yum: pkg=mariadb state=latest
```

像上面例子中针对每一组主机的所有操作就组成一个 Play，一般一个 Playbook 中只包含一个 Play，所以不用太在意如何区分 Playbook 与 Play。当在一些 Ansible 文档中提到 Play 的概念时，知道是怎么回事就可以了。

2.5 Ansible 模块

2.5.1 什么是 Ansible 模块

bash 无论是在命令行上执行，还是在 bash 脚本中，都需要调用 cd、ls、copy、yum 等命令。模块就是 Ansible 的"命令"。模块是 Ansible 命令行和脚本中都需要调用的。常用的 Ansible 模块有 yum、copy、template 等。

bash 在调用命令时可以跟不同的参数，每个命令的参数都是该命令自定义的。同样，Ansible 中调用模块也可以跟不同的参数，每个模块的参数也都是由模块自定义的。

模块的详细用法可以查阅下面的文档，也可以通过命令"ansible-doc <module_name>"查看其用法：

http://docs.ansible.com/ansible/modules_by_category.html

2.5.2　在命令行里使用模块

在命令行中：

◎　-m 后面接调用模块的名字。
◎　-a 后面接调用模块的参数。

```
$ #使用模块 copy 复制管理员节点文件/etc/hosts 到所有远程主机/tmp/hosts
$ ansible all -m copy -a "src=/etc/hosts dest=/tmp/hosts"
$ #使用模块 yum 在远程主机 Web 上安装 httpd 包
$ ansible web -m yum -a "name=httpd state=present"
```

2.5.3　在 Playbook 脚本中使用模块

在 Playbook 脚本中，tasks 中的每一个 Action 都是对模块的一次调用。在每个 Action 中：

◎　冒号前面是模块的名字。
◎　冒号后面是调用模块的参数。

```
---
 tasks:
 - name: ensure apache is at the latest version
   yum: pkg=httpd state=latest
 - name: write the apache config file
   template: src=templates/httpd.conf.j2 dest=/etc/httpd/conf/httpd.conf
 - name: ensure apache is running
   service: name=httpd state=started
```

2.5.4　Ansible 模块的特点

❶ 像 Linux 中的命令一样，Ansible 的模块既可以在命令行中调用，也可以在 Ansible 的脚本 Playbook 中调用。

❷ 每个模块的参数和状态的判断，都取决于该模块的具体实现，所以在使用它们之前需要查阅该模块对应的文档。

❸ 可以通过文档查看其具体的用法（http://docs.ansible.com/ansible/list_of_all_modules.html）。

❹ 通过命令 ansible-doc 也可以查看模块的用法。

❺ Ansible 提供一些常用功能的模块，同时 Ansible 也提供 API，让用户可以自己写模块，使用的编程语言是 Python。

2.5.5 常用模块

学习 Ansible 时，非常有必要了解一些常用的模块。

接下来介绍一些后面章节中会用到的模块，也是很常用的模块。

❶ 调试和测试类的模块。

- ◎ ping：ping 一下你的远程主机，如果可以通过 Ansible 连接成功，那么返回 pong。
- ◎ debug：用于调试的模块，只是简单打印一些消息，有点像 Linux 的 echo 命令。

❷ 文件类的模块。

- ◎ copy：从本地复制文件到远程节点。
- ◎ template：从本地复制文件到远程节点，并进行变量的替换。
- ◎ file：设置文件属性。

❸ Linux 上的常用操作。

- ◎ user：管理用户账户。
- ◎ yum：Red Hat 系 Linux 上的包管理。
- ◎ service：管理服务。
- ◎ firewalld：管理防火墙中的服务和端口。

❹ 执行 shell 命令。

- ◎ shell：在节点上执行 shell 命令，支持$HOME、"<"、">"、"|"、";" 和 "&"。
- ◎ command：在远程节点上面执行命令，不支持$HOME、"<"、">"、"|"、";" 和 "&"。

1. ping 模块

这个是测试远程节点的 SSH 连接是否就绪的常用模块。但它并不像 Linux 命令那样简单地 ping 一下远程节点，而是先检查能否通过 SSH 登录远程节点，再检查其 Python 版本能否满足要求，如果都满足则会返回 pong，表示成功。

ping 无须任何参数。因为 ping 是测试节点连接的可用性,所以通常在命令行中使用得较为频繁。下面是 ping 在命令行中的用法。

```
ansible servers -m ping
```

2. debug 模块

打印输出信息,类似 Linux 上的 echo 命令。

❶ 通过参数 msg 定义打印的字符串。

msg 中可以嵌入变量,下面的例子中注入了系统变量,Ansible 在执行 Playbook 之前会收集一些比较常用的系统变量,这样在 Playbook 中不需要定义直接使用就可以。

```
- debug:
    msg: "System {{ inventory_hostname }} has gateway {{ ansible_default_ipv4.gateway }}"
```

执行结果:

```
TASK [debug] ****************************************************************
ok: [localhost] => {
    "msg": "System localhost has gateway 192.168.50.1"
}
```

❷ 通过参数 var 定义需要打印的变量。

变量可以是系统变量,也可以是动态的执行结果,通过关键字 regester 注入变量中。

◎ 打印系统变量。

```
- name: Display all variables/facts known for a host
  debug:
    var: hostvars[inventory_hostname]["ansible_default_ipv4"]["gateway"]
```

执行结果:

```
    TASK [Display part of variables/facts known for a host] **********************
    ok: [localhost] => {
        "hostvars[inventory_hostname][\"ansible_default_ipv4\"][\"gateway\"]": "192.168.50.1"
    }
```

◎ 打印动态注入的变量。

```
- shell: /usr/bin/uptime
  register: result

- debug:
    var: result
```

执行结果：

```
   TASK command ***************************************************
***** changed: [localhost]
   TASK debug *****************************************************
***** ok: [localhost] => { "result": { "changed": true,
        "cmd": "/usr/bin/uptime",
        "delta": "0:00:00.003212",
        "end": "2017-01-01 21:30:02.817443",
        "rc": 0,
        "start": "2017-01-01 21:30:02.814231",
        "stderr": "",
        "stdout": " 21:30:02 up 12:38,  8 users,  load average: 1.13, 1.31, 1.14",
        "stdout_lines": [
            " 21:30:02 up 12:38,  8 users,  load average: 1.13, 1.31, 1.14"
        ],
        "warnings": []
    }
}
```

3. copy 模块

从当前的机器上复制静态文件到远程节点上，并且设置合理的文件权限。注意，copy 模块在复制文件的时候，会先比较一下文件的 checksum，如果相同则不会复制，返回状态为 OK；如果不同才会复制，返回状态为 changed。

❶ 设置文件权限。

利用 mode 设置权限可以是数字，当然也可以是符号的形式"u=rw,g=r,o=r"和"u+rw, g-wx, o-rwx"。

```
- copy:
    src: /srv/myfiles/foo.conf
    dest: /etc/foo.conf
    owner: foo
    group: foo
    mode: 0644
```

❷ 备份节点上原来的文件。

backup 参数为 yes 的时候，如果发生了复制（copy）操作，那么会先复制目标节点上的源文件。当两个文件相同时，不再进行复制操作。

```
- copy:
    src: sudoers
    dest: /tmp
    backup: yes
```

❸ 复制后的验证操作。

validate 参数接需要验证的命令。一般需要验证复制后的文件，所以%s 都可以指代复制后的文件。当 copy 模块中加入了 validate 参数后，不仅需要成功复制文件，还需要 validate 命令返回成功的状态，整个模块的执行状态才算成功。

```
visudo -cf /etc/sudoers
```

上面的命令是验证 sudoers 文件有没有语法错误的命令。

```
- copy:
    src: /mine/sudoers
    dest: /etc/sudoers
    validate: 'visudo -cf %s'
```

4. template 模块

如果复制的只是静态文件，那么用 copy 模块就足够了。但如果在复制的同时需要根据实际情况修改部分内容，那么就需要用到 template 模块。

比如安装 Apache 后，你需要给节点复制一个测试页面 index.html。index.html 里面需要显示当前节点的主机名和 IP，这时候就需要用到 template。

在 index.html 中，需要指定想替换的是哪个部分，那么这个部分就用变量来表示。template

使用的是 Python 的 Jinja2 模板引擎。这里读者不需要了解 Jinja2，只需要知道变量的表示法是 **{{}}** 就可以了。

❶ template 文件语法。

index.html 具体应该怎么写呢？既然是 template 文件，那么我们就加一个后缀来提高可读性，index.html.j2。下面文件中使用了两个变量，ansible_hostname 和 ansible_default_ipv4.address。

```
<html>
<title>Demo</title>
<body>
<div class="block" style="height: 99%;">
    <div class="centered">
        <h1>#46 Demo</h1>
        <p>Served by {{ ansible_hostname }} ({{ ansible_default_ipv4.address }}).</p>
    </div>
</div>
</body>
</html>
```

❷ 使用 Facts 变量（远程主机的系统变量）的 template。

index.html.j2 使用的两个变量 ansible_hostname 和 ansible_default_ipv4.address 都是远程主机的系统变量，Ansible 会替我们搜索，可以直接在 Playbook 中使用，当然也可以直接在 template 中使用。所以我们在写 template 语句时无须传入参数。

```
- name: Write the default index.html file
  template: src=templates/index.html.j2 dest=/var/www/html/index.html
```

❸ 使用普通变量的 template。

将 httpd.conf.j2 复制到远程节点，根据需求设置默认的 HTTP 端口，这时我们就需要用到普通的变量。

在 httpd.conf.j2 模板文件中，所有变量的用法都是一样的，都是用 **{{}}**。

```
ServerRoot "/etc/httpd"
...
Listen {{ http_port }}
...
```

普通变量不是在调用 template 的时候传进去，而是通过 Playbook 中的 vars 关键字定义。当然，在 Playbook 中可以直接使用的变量，都可以在 template 中直接使用，包括后面章节会提到的定义在 inventory 中的变量。

```
- hosts: localhost
  vars:
    http_port: 8080
    remote_user: root

  tasks:
  - name: Write the configuration file
    template: src=templates/httpd.conf.j2 dest=/etc/httpd/conf/httpd.conf
```

❹ 和 copy 模块一样强大的功能。

copy 模块不仅可以简单地复制文件到远程节点，还可以进行权限设置、文件备份，以及验证功能，这些功能 template 同样具备。

```
- template:
    src: etc/ssh/sshd_config.j2
    dest: /etc/ssh/sshd_config.j2
    owner: root
    group: root
    mode: '0600'
    validate: /usr/sbin/sshd -t %s
    backup: yes
```

5. file 模块

file 模块可以用来设置远程主机上的文件、软链接（symlinks）和文件夹的权限，也可以用来创建和删除它们。

❶ 改变文件的权限。

mode 参数既可以直接赋值数字权限（必须以 0 开头），还可以用来增加和删除权限。具体的写法见下面的代码。

```
- file:
    path: /etc/foo.conf
    owner: foo
```

```yaml
    group: foo
    mode: 0644
    #mode: "u=rw,g=r,o=r"
    #mode: "u+rw,g-wx,o-rwx"
```

❷ 创建文件的软链接。

注意，这里面的 src 和 dest 参数的含义和 copy 模块中的不一样，file 模块里面所操作的文件都是远程节点上的文件。

```yaml
- file:
    src: /file/to/link/to
    dest: /path/to/symlink
    owner: foo
    group: foo
    state: link
```

❸ 创建一个新文件。

像 touch 命令一样创建一个新文件。

```yaml
- file:
    path: /etc/foo.conf
    state: touch
    mode: "u=rw,g=r,o=r"
```

❹ 创建新的文件夹。

```yaml
# create a directory if it doesn't exist
- file:
    path: /etc/some_directory
    state: directory
    mode: 0755
```

6. user 模块

user 模块可以增、删、改 Linux 远程节点的用户账户，并为其设置账户的属性。

❶ 增加账户。

◎ 增加账户 johnd，并且设置 uid 为 1040，设置用户的 primary group 为 admin：

```yaml
- user:
```

```yaml
    name: johnd
    comment: "John Doe"
    uid: 1040
    group: admin
```

- 创建账户 james，并将其添加到两个 group 中。

```yaml
- user:
    name: james
    shell: /bin/bash
    groups: admins,developers
    append: yes
```

❷ 删除账户。

删除账户 johnd。

```yaml
- user:
    name: johnd
    state: absent
    remove: yes
```

❸ 修改账户的属性。

- 为账户 jsmith 创建一个 2048 位的 SSH 密钥，放在 ~jsmith/.ssh/id_rsa 中。

```yaml
- user:
    name: jsmith
    generate_ssh_key: yes
    ssh_key_bits: 2048
    ssh_key_file: .ssh/id_rsa
```

- 为用户添加过期时间。

```yaml
- user:
    name: james18
    shell: /bin/zsh
    groups: developers
    expires: 1422403387
```

7. yum 模块

yum 模块是用来管理 Red Hat 系的 Linux 上的安装包的，包括 RHEL、CentOS 和

Fedora 21 及以下版本。Fedora 从版本 22 开始就使用 dnf，推荐使用 dnf 模块来进行安装包的操作。

❶ 从 yum 源上安装和删除包。

◎ 安装最新版本的包，如果已经安装了老版本，那么会更新到最新的版本。

```
- name: install the latest version of Apache
  yum:
    name: httpd
    state: latest
```

◎ 安装指定版本的包。

```
- name: install one specific version of Apache
  yum:
    name: httpd-2.2.29-1.4.amzn1
    state: present
```

◎ 删除 httpd 包。

```
- name: remove the Apache package
  yum:
    name: httpd
    state: absent
```

◎ 从指定的 repo testing 中安装包。

```
- name: install the latest version of Apache from the testing repo
  yum:
    name: httpd
    enablerepo: testing
    state: present
```

❷ 从 yum 源上安装一组包。

```
- name: install the 'Development tools' package group
  yum:
    name: "@Development tools"
    state: present

- name: install the 'Gnome desktop' environment group
  yum:
```

```yaml
    name: "@gnome-desktop-environment"
    state: present
```

❸ 从本地文件中安装包。

```yaml
- name: install nginx rpm from a local file
  yum:
    name: /usr/local/src/nginx-release-centos-6-0.el6.ngx.noarch.rpm
    state: present
```

❹ 从 URL 中安装包。

```yaml
- name: install the nginx rpm from a remote repo
  yum:
    name: http://nginx.org/packages/centos/6/noarch/RPMS/nginx-release-centos-6-0.el6.ngx.noarch.rpm
    state: present
```

8. service 服务管理模块

该模块用来管理远程节点上的服务，比如 httpd、sshd、nfs、crond 等。

❶ 开、关、重启、重载服务。

◎ 开启服务。

```yaml
- service:
    name: httpd
    state: started
```

◎ 关服务。

```yaml
- service:
    name: httpd
    state: stopped
```

◎ 重启服务。

```yaml
- service:
    name: httpd
    state: restarted
```

◎ 重载服务。

```
- service:
    name: httpd
    state: reloaded
```

❷ 设置开机启动的服务。

```
- service:
    name: httpd
    enabled: yes
```

❸ 启动网络服务下的接口 eth0。

```
- service:
    name: network
    state: restarted
    args: eth0
```

9. firewalld 模块

firewalld 模块为某服务和端口添加 firewalld 规则。firewalld 中有正在运行的规则和永久的规则，firewalld 模块都支持。

firewalld 要求远程节点上的 firewalld 版本在 0.2.11 以上。

❶ 为服务添加 firewalld 规则。

```
- firewalld:
    service: https
    permanent: true
    state: enabled
```

```
- firewalld:
    zone: dmz
    service: http
    permanent: true
    state: enabled
```

❷ 为端口号添加 firewalld 规则。

```
- firewalld:
    port: 8081/tcp
```

```yaml
      permanent: true
      state: disabled
  - firewalld:
      port: 161-162/udp
      permanent: true
      state: enabled
```

❸ 其他复杂的 firewalld 规则。

```yaml
  - firewalld:
      rich_rule: 'rule service name="ftp" audit limit value="1/m" accept'
      permanent: true
      state: enabled
  - firewalld:
      source: 192.0.2.0/24
      zone: internal
      state: enabled
  - firewalld:
      zone: trusted
      interface: eth2
      permanent: true
      state: enabled
  - firewalld:
      masquerade: yes
      state: enabled
      permanent: true
      zone: dmz
```

10. shell 模块

在远程节点上通过/bin/sh 执行命令。如果一个操作可以通过模块 yum、copy 实现，那么建议不要使用 shell 或者 command 这样通用的命令模块。因为通用的命令模块不会根据具体操作的特点进行状态（status）判断，所以当没有必要再重新执行的时候，它还是会重新执行一遍。

❶ 支持 HOME、"<"、">"、"|"、";" 和 "&"。

◎ 支持$home。

```yaml
- name: test $home
```

```
    shell: echo "Test1" > ~/tmp/test1
```

- ◎ 支持"&&"。

```
- shell: service jboss start && chkconfig jboss on
```

- ◎ 支持">>"。

```
- shell: echo foo >> /tmp/testfoo
```

❷ 调用脚本。

- ◎ 调用脚本。

```
- shell: somescript.sh >> somelog.txt
```

- ◎ 在执行命令前改变工作目录。

```
- shell: somescript.sh >> somelog.txt
  args:
    chdir: somedir/
```

- ◎ 在执行命令前改变工作目录，并且仅在文件 somelog.txt 不存在时执行命令。

```
- shell: somescript.sh >> somelog.txt
  args:
    chdir: somedir/
    creates: somelog.txt
```

- ◎ 指定用 bash 运行命令。

```
- shell: cat < /tmp/\*txt
  args:
    executable: /bin/bash
```

11. command 模块

在远程节点上执行命令。和 shell 模块类似，但不支持$HOME "<"、">"、"|"、";"和"&"等操作。

❶ 与 shell 模块相同之处。

- ◎ 都可以调用单条命令。

```
- command: /sbin/shutdown -t now
```

◎ 都可以在执行命令前改变目录,并仅在某个文件(例如 database)不存在时再执行。

```
- command: /usr/bin/make_database.sh arg1 arg2
  args:
    chdir: somedir/
    creates: /path/to/database
```

❷ 和 shell 模块不同之处。

◎ command 模块多了一个传参方式。

```
- command: /usr/bin/make_database.sh arg1 arg2 creates=/path/to/database
```

◎ command 模块不支持 "&&" 和 ">>"。

下面的写法是无法创建/tmp/test3 和/tmp/test4 的。

```
- name: test $home
  command: echo "test3" > ~/tmp/test3 && echo "test4" > ~/tmp/test4
```

第 3 章

Ansible 进阶

本章重点

3.1 Ansible 的配置
3.2 主机清单
3.3 Ansible 的脚本 Playbook
3.4 更多的 Ansible 模块
3.5 最佳使用方法

3.1 Ansible 的配置

3.1.1 可以配置什么

Ansible 可以配置主机清单文件"inventory"、extra 模块放置路径"library"、远程主机的临时文件位置"remote_tmp",以及管理节点上临时文件的位置"local_tmp"。

```
inventory       = /etc/ansible/hosts
library         = /usr/share/my_modules/
remote_tmp      = $HOME/.ansible/tmp
local_tmp       = $HOME/.ansible/tmp
```

还可以配置连接端口号"accelerate_port"、超时时间等。

```
accelerate_port = 5099
accelerate_timeout = 30
accelerate_connect_timeout = 5.0
```

安装好 Ansible 后,通过/etc/ansible/ansible.cfg 文件的内容和注释就可以了解到所有可以配置的选项,也可以通过下面的链接在线查看 ansible.cfg 文件。

https://raw.githubusercontent.com/ansible/ansible/devel/examples/ansible.cfg

如果对 Ansible 配置文件里面的关键词不能完整理解,还可以参考关键词解释列表:

http://docs.ansible.com/ansible/intro_configuration.html#explanation-of-values-by-section

3.1.2 Ansible 配置文件的优先级

Ansible 的默认配置文件是/etc/ansible/ansible.cfg。其实 Ansible 会按照下面的顺序查找配置文件,并使用第一个发现的配置文件。

```
* ANSIBLE_CONFIG (an environment variable)
* ansible.cfg (in the current directory)
* .ansible.cfg (in the home directory)
* /etc/ansible/ansible.cfg
```

Ansible 1.5 以前版本的顺序如下。

```
* ansible.cfg (in the current directory)
* ANSIBLE_CONFIG (an environment variable)
* .ansible.cfg (in the home directory)
* /etc/ansible/ansible.cfg
```

3.2 主机清单

什么是主机清单（Host Inventory）呢？它是告诉 Ansible 需要管理哪些主机，以及这些主机的分类和分组信息的文件。在实际使用中，既可根据远程主机所在的地域分类，也可按照其功能来分类。

❶ Inventory 的配置文件。

默认文件：

```
/etc/ansible/hosts
```

❷ 在配置文件中修改 Inventory 文件的位置。

在文件/etc/ansible/ansible.cfg 中修改。

```
...
inventory       = /etc/ansible/hosts
...
```

❸ 命令行中传递 Inventory 配置文件。

◎ 利用参数-i 传递主机清单配置文件。

```
$ ansible-playbook -i hosts site.yml
```

◎ 利用参数--inventory-file。

```
$ ansible-playbook --inventory-file hosts site.yml
```

3.2.1 远程主机的分组

给远程主机分组,以便于在 Playbook 中使用。下面的文件中,展示了主机清单文件中最简单的分组方法,[]内是组名。将远程主机分为 webservers、dbservers、databases 几个组。

```
mail.example.com

[dbservers]
one.example.com
two.example.com
three.example.com

[webservers]
www[01:50].example.com

[databases]
db-[a:f].example.com
```

当然分组也可以支持嵌套。在下面的例子中,usa 组还可以包含其他的组,例如[usa]中包含 southeast 组,southeast 组中还可以包含组 atlanta 和 releigh。

```
[atlanta]
host1
host2

[raleigh]
host2
host3

[southeast:children]
atlanta
raleigh

[usa:children]
southeast
northeast
southwest
northwest
```

3.2.2 设置连接参数

Ansible 可以在 Inventory 文件中指定主机的连接参数，包括连接方法、用户等。在 Inventory 中设置连接的参数如下，用空格分隔多个参数。

```
[targets]

localhost              ansible_connection=local
other1.example.com     ansible_connection=ssh   ansible_user=root ansible_user=mpdehaan
other2.example.com     ansible_connection=ssh   ansible_user=mdehaan
```

其他常用的连接参数如表 3.1 所示。

表 3.1 常用的连接参考数

连接参数的值	连接参数的含义
ansible_connection	SSH 的连接方式。可以指定为 smart、ssh 或者 paramiko
ansible_host	Ansible 连接的主机地址，如果你在 Ansible 中给主机起了一个不同的别名，那么需要用这个参数
ansible_port	SSH 端口号，默认为 22
ansible_user	SSH 连接时使用的默认用户名
ansible_ssh_pass	SSH 连接时使用的密码。不过不建议用本参数存储明文的密码，尽量使用 values 对密码进行加密存储
ansible_ssh_private_key_file	基于 key 的 SSH 连接，使用的是 private key 文件
ansible_ssh_common_args	通过配置本参数来指定 SFTP、SCP 和 SSH 默认的额外参数

所有可以指定的参数在下列文档中：

http://docs.ansible.com/ansible/intro_inventory.html#list-of-behavioral-inventory-parameters

3.2.3 变量

Ansible 支持在主机清单文件中置变量，或与主机清单文件同目录的特定子目录和文件中定义变量。下面以设置 NTP 服务器、代理服务器和数据库地址为例。

1. 主机清单文件中的变量

◎ 为单个远程主机指定参数。

```
[atlanta]
host1 http_port=80 maxRequestsPerChild=808
host2 http_port=303 maxRequestsPerChild=909
```

◎ 为一个组指定变量。

```
[atlanta]
host1
host2

[atlanta:vars]
ntp_server=ntp.atlanta.example.com
proxy=proxy.atlanta.example.com
```

2. 按目录结构存储变量

假设主机清单文件为 /etc/ansible/hosts，那么相关的 Host 和 Group 变量可以放在 /etc/ansible/host_vars/和/etc/ansible/group_vars/下的主机同名目录中的文件，也支持 '.yml'、'.yaml' 和'.json'为后缀的 YMAL 和 JSON 文件，如下面的例子所示。

```
/etc/ansible/group_vars/raleigh #变量的文件名还可以是 '.yml'、'.yaml'和
'.json'
/etc/ansible/group_vars/webservers
/etc/ansible/host_vars/foosball
```

/etc/ansible/group_vars/raleigh 文件内容可以为。

```
---
ntp_server: acme.example.org
database_server: storage.example.org
```

如果对应的名字为目录名，则 Ansible 会读取这个目录下面所有文件的内容。在下面的例子中，db_settings 和 cluster_settings 中的变量都会被 Ansible 读取。

```
/etc/ansible/group_vars/raleigh/db_settings
/etc/ansible/group_vars/raleigh/cluster_settings
```

group_vars/和 host_vars/目录可放在 Inventory 文件同级目录下，或是 Playbook 文件同级目录下。如果两个目录下都存在变量文件，那么 Playbook 目录下的值会覆盖 Inventory 目录下变量的值。

3.3 Ansible 的脚本 Playbook

3.3.1 Playbook 的文件格式 YAML

Playbook 是 Ansible 的脚本语言，使用的是 YAML 格式。YAML 和 JSON 类似，是一种数据表示格式，下面介绍一些关于 YAML 语言的基本知识。

❶ 文件开始符。

```
---
```

❷ 数组 List。

```
- element1
- element2
- element3
```

数组中的每一个元素都是以 - 开始的。

❸ 字典（Hash or Directory）。

```
key: value
```

key 和 value 之间用冒号加空格分隔。

❹ 复杂的字典。

字典的嵌套。

```
# An employee record
martin:
    name: Martin D'vloper
    job: Developer
    skill: Elite
```

字典和数组的嵌套。

```
- martin:
    name: Martin D'vloper
    job: Developer
```

```
    skills:
      - python
      - perl
      - pascal
- tabitha:
    name: Tabitha Bitumen
    job: Developer
    skills:
      - lisp
      - fortran
      - erlang
```

❺ 需要注意的地方。

变量里有冒号（:）时要加引号。

```
foo: "somebody said I should put a colon here: so I did"
```

变量以"{"开头时要加引号。

```
foo: "{{ variable }}"
```

3.3.2 执行 Playbook 的命令

如何执行 Ansible 的脚本 Playbook 呢？Ansible 提供了一个单独的命令：ansible-playbook，常见的 ansible-playbook 的使用方法如下。

◎ 执行 Playbook 的基本方法。

```
$ ansible-playbook deploy.yml
```

◎ 查看输出的细节。

```
ansible-playbook playbook.yml --verbose
```

◎ 查看该脚本影响哪些主机（hosts）。

```
ansible-playbook playbook.yml --list-hosts
```

◎ 并行执行脚本。

```
ansible-playbook playbook.yml -f 10
```

更多的关于 ansible-playbook 的用法请查看"man ansible-playbook"或者"ansible-playbook -h"。

3.3.3 Playbook 的基本语法

最基本的 Playbook 脚本分为三个部分。

❶ 在什么机器上以什么身份执行。

◎ hosts
◎ users
◎ ...

❷ 执行的任务是都有什么。

◎ tasks

❸ 善后的任务都有什么。

◎ handlers

下面展示了一个完整的 Playbook 的示例。

```
---
- hosts: webservers
  user: root
  vars:
    http_port: 80
    max_clients: 200

  tasks:
  - name: ensure apache is at the latest version
    yum: pkg=httpd state=latest
  - name: write the apache config file
    template: src=/srv/httpd.j2 dest=/etc/httpd.conf
    notify:
    - restart apache
  - name: ensure apache is running
    service: name=httpd state=started
  handlers:
    - name: restart apache
      service: name=httpd state=restarted
```

1. 主机和用户

主机和用户如表 3.2 所示。

表 3.2 主机和用户

key	含 义
hosts	为主机的 IP，或者主机组名，或者关键字 all
user	在远程以哪个用户身份执行
become	切换成其他用户身份执行，值为 yes 或者 no
become_method	与 became 一起用，指可以为 'sudo'/'su'/'pbrun'/'pfexec'/'doas'
become_user	与 became_user 一起用，可以是 root 或者其他用户名

脚本里用 become 时，执行的 Playbook 必须加参数--ask-become-pass，提示用户输入"sudo"的密码。

```
ansible-playbook deploy.yml --ask-become-pass
```

2. 任务列表

- ◎ 任务（task）是从上至下顺序执行的，如果中间发生错误，那么整个 Playbook 会中止。但可以改修文件后，再重新执行。
- ◎ 每一个任务都是对模块的一次调用，只是使用不同的参数和变量而已。
- ◎ 每一个任务最好有 name 属性，这是供人读的，没有实际的操作。然后会在命令行里面输出，提示用户执行情况。

❶ 基本语法。

任务的基本写法。

```
tasks:
  - name: make sure apache is running
    service: name=httpd state=running
```

其中 name 是可选的，也可以简写成下面的样子。

```
tasks:
  - service: name=httpd state=running
```

写了 name 的任务在 Playbook 执行时，会显示对应的名字，信息更友好、丰富。写 name 是个好习惯。

```
TASK: [make sure apache is running] ********************************
****************************
changed: [yourhost]
```

没有写 name 的任务在 Playbook 执行时，直接显示对应的任务语法。在多次调用同样的模块后，不容易分辨执行到哪一步了。

```
TASK: [service name=httpd state=running] ****************************
***********
changed: [yourhost]
```

❷ 参数的不同写法。

最上面的代码展示了最基本的传入模块的参数的方法 key=value。

```
tasks:
  - name: make sure apache is running
    service: name=httpd state=running
```

当需要传入参数列表过长时，可以分隔到多行。

```
tasks:
  - name: Copy ansible inventory file to client
    copy: src=/etc/ansible/hosts dest=/etc/ansible/hosts
          owner=root group=root mode=0644
```

或者用 YML 的字典格式传入参数。

```
tasks:
  - name: Copy ansible inventory file to client
    copy:
      src: /etc/ansible/hosts
      dest: /etc/ansible/hosts
      owner: root
      group: root
      mode: 0644
```

❸ 任务的执行状态。

任务中每个 Action 会调用一个模块，然后在模块中检查当前系统状态是否需要重新执行。

◎ 如果本次执行了，那么 Action 会得到返回值 changed。

◎ 如果不需要执行，那么 Action 会得到返回值 ok。

模块的执行状态的具体判断规则由各个模块自己决定和实现。例如，copy 模块的判断方法是比较文件的 checksum，代码如下。

```
checksum_src = module.sha1(src)
...
checksum_dest = module.sha1(dest)
...
if checksum_src != checksum_dest or os.path.islink(b_dest):
    ...
    changed = True
else:
    changed = False
```

完整的 copy 模块的代码请参考下面的链接：

https://github.com/ansible/ansible-modules-core/blob/devel/files/copy.py

下面以一个 copy 文件的任务为例，展示任务状态在执行时到底有什么不同的行为。

```
tasks:
- name: Copy the /etc/hosts
  copy: src=/etc/hosts dest=/etc/hosts
```

第一次执行，它的结果如图 3.1 所示。

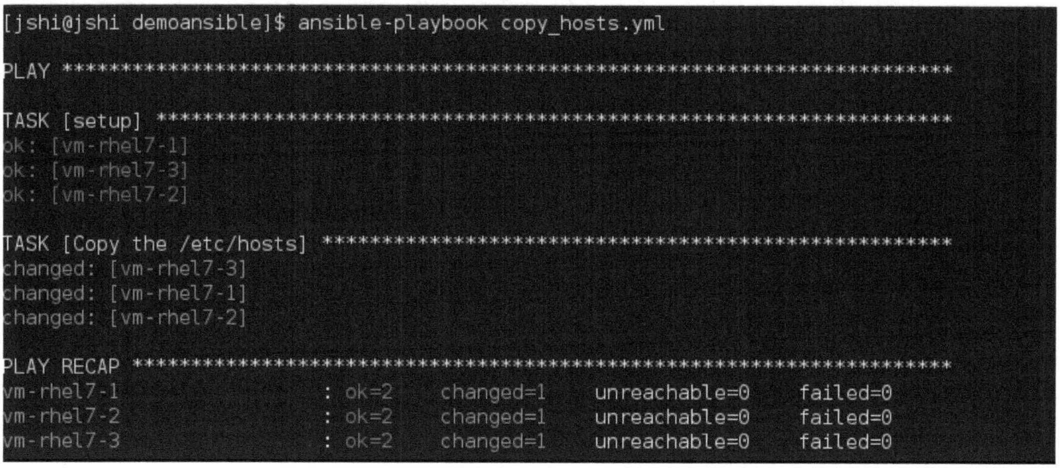

图 3.1　任务的状态是 changed

第二次执行时它的结果如图 3.2 所示。

```
[jshi@jshi demoansible]$ ansible-playbook copy_hosts.yml

PLAY ****************************************************************

TASK [setup] ********************************************************
ok: [vm-rhel7-3]
ok: [vm-rhel7-1]
ok: [vm-rhel7-2]

TASK [Copy the /etc/hosts] ******************************************
ok: [vm-rhel7-1]
ok: [vm-rhel7-3]
ok: [vm-rhel7-2]

PLAY RECAP **********************************************************
vm-rhel7-1                 : ok=2    changed=0    unreachable=0    failed=0
vm-rhel7-2                 : ok=2    changed=0    unreachable=0    failed=0
vm-rhel7-3                 : ok=2    changed=0    unreachable=0    failed=0
```

图 3.2　任务的状态是 ok

由于第一次执行 copy_hosts.yml 时，已经复制过文件，因此 Ansible 会根据目标文件的状态避免重复执行复制。

接着更改 vm-rhel7-1 的 /etc/hosts，再次执行看看。如图 3.3 所示，只有 vm-rhel7-1 的状态是 changed，其他两台远程主机的状态是 ok。

```
[jshi@jshi demoansible]$ ansible-playbook copy_hosts.yml

PLAY ****************************************************************

TASK [setup] ********************************************************
ok: [vm-rhel7-1]
ok: [vm-rhel7-3]
ok: [vm-rhel7-2]

TASK [Copy the /etc/hosts] ******************************************
ok: [vm-rhel7-2]
ok: [vm-rhel7-3]
changed: [vm-rhel7-1]

PLAY RECAP **********************************************************
vm-rhel7-1                 : ok=2    changed=1    unreachable=0    failed=0
vm-rhel7-2                 : ok=2    changed=0    unreachable=0    failed=0
vm-rhel7-3                 : ok=2    changed=0    unreachable=0    failed=0
```

图 3.3　主机的状态

3. 响应事件 handler

❶ 什么是 handler

每个主流的编程语言都有 Event 机制，而 handler 就是 Playbook 的 Event。

Handlers 里面的每一个 handler 都是对模块的一次调用。而 handler 与任务不同，任务会默认地按定义顺序执行每一个任务，handler 则不会，它需要在任务中被调用，才有可能被执行。

前面提过，任务表中的任务都是有状态的：changed 或者 ok。在 Ansible 中，只有在任务的执行状态为 changed 时，才会执行该任务调用的 handler。这也是 handler 与普通的 Event 机制不同的地方。

❷ 应用场景

什么情况下使用 handler 呢？

如果你在任务中修改了 Apache 的配置文件，那么需要重启 Apache。如果还安装了 Apache 的插件，那么还需要重启 Apache。像这样的应用场景中，重启 Apache 就可以设计成一个 handler。

一个 handler 最多只执行一次，并且在所有的任务都执行完之后再执行。如果有多个任务调用（notify）同一个 handler，那么只执行一次。

在下面的例子中 Apache 只执行一次。

```
---
- hosts: lb
  remote_user: root
  vars:
     random_number1: "{{ 10000 | random }}"
     random_number2: "{{ 10000000000 | random }}"
  tasks:
  - name: Copy the /etc/hosts to /tmp/hosts.{{ random_number1 }}
    copy: src=/etc/hosts dest=/tmp/hosts.{{ random_number1 }}
    notify:
       - call in every action
  - name: Copy the /etc/hosts to /tmp/hosts.{{ random_number2 }}
    copy: src=/etc/hosts dest=/tmp/hosts.{{ random_number2 }}
    notify:
       - call in every action
```

```
handlers:
- name: call in every action
  debug: msg="call in every action, but execute only one time"
```

只有 changed 状态的任务才会触发 handler 的执行。

当任务的执行状态为 changed 时，才会触发 notify handler 的执行。

下面的脚本执行两次，执行结果是不同的。

◎ 第一次执行时：
— 任务的状态都是 changed，会触发两个 handler。
◎ 第二次执行时：
— 第一个任务的状态是 ok，因而不会触发 handlers"call by /tmp/hosts"。
— 第二个任务的状态是 changed，触发了 handler "call by /tmp/hosts.random_number"。

测试代码如下。

```
---
- hosts: lb
  remote_user: root
  vars:
      random_number: "{{ 10000 | random }}"
  tasks:
  - name: Copy the /etc/hosts to /tmp/hosts
    copy: src=/etc/hosts dest=/tmp/hosts
    notify:
      - call by /tmp/hosts
  - name: Copy the /etc/hosts to /tmp/hosts.{{ random_number }}
    copy: src=/etc/hosts dest=/tmp/hosts.{{ random_number }}
    notify:
      - call by /tmp/hosts.random_number

  handlers:
  - name: call by /tmp/hosts
    debug: msg="call first time"
  - name: call by /tmp/hosts.random_number
    debug: msg="call by /tmp/hosts.random_number"
```

❸ handler 按定义的顺序执行。

handler 是按照定义的顺序执行的,而不是按照安装在任务中调用的顺序执行的。

下面的例子定义的顺序是 1>2>3,调用的顺序是 3>2>1,实际执行顺序:1>2>3。

```
---
- hosts: lb
  remote_user: root
  gather_facts: no
  vars:
      random_number1: "{{ 10000 | random }}"
      random_number2: "{{ 10000000000 | random }}"
  tasks:
  - name: Copy the /etc/hosts to /tmp/hosts.{{ random_number1 }}
    copy: src=/etc/hosts dest=/tmp/hosts.{{ random_number1 }}
    notify:
      - define the 3nd handler
  - name: Copy the /etc/hosts to /tmp/hosts.{{ random_number2 }}
    copy: src=/etc/hosts dest=/tmp/hosts.{{ random_number2 }}
    notify:
      - define the 2nd handler
      - define the 1nd handler

  handlers:
  - name: define the 1nd handler
    debug: msg="define the 1nd handler"
  - name: define the 2nd handler
    debug: msg="define the 2nd handler"
  - name: define the 3nd handler
    debug: msg="define the 3nd handler"
```

3.3.4 变量

在本节,我们主要介绍几种常用的变量,在后面的章节中,我们会单独介绍一些复杂情景下变量的使用和覆盖原则。

在 Playbook 中,常用的几种变量包括以下几种情况:

❶ 在 Playbook 中用户自定义的变量。

❷ 用户无须自定义，Ansible 会在执行 Playbook 之前去全程主机上搜集关于远程节点系统信息的变量。

❸ 在文件模板中，可以直接使用上述两种变量。

❹ 把任务的运行结果作为一个变量来使用，这个变量叫作注册变量。

❺ 为了使 Playbook 更灵活、通用性更强，允许用户在执行 Playbook 时传入变量的值，这个时候就需要用到"额外变量"。

1. 在 Playbook 中用户自定义的变量

用户可以在 Playbook 中，通过 vars 关键字自定义变量，使用时用{{ }}引用起来即可。

❶ Playbook 中定义和使用变量的方法

例如，下面的例子中，用户定义变量名为 http_port，其值为 80。在 tasks 下的 firewalld 中，通过 {{ http_port }} 引用：

```
---
- hosts: web
  vars:
    http_port: 80
  remote_user: root
  tasks:
  - name: insert firewalld rule for httpd
    firewalld: port={{ http_port }}/tcp permanent=true state=enabled immediate=yes
```

❷ 把变量放在单独的文件中

当变量较多的时候，或者变量需要在多个 Playbook 中重用的时候，可以把变量放到一个单独的文件中，通过关键字 "var_files" 把文件中定义的变量引用到 Playbook 中。使用变量的方法和在本文件中定义变量的使用方法相同。

```
- hosts: web
  remote_user: root
  vars_files:
     - vars/server_vars.yml
```

```
    tasks:
    - name: insert firewalld rule for httpd
      firewalld: port={{ http_port }}/tcp permanent=true state=enabled immediate=yes
```

变量文件 vars/server_vars.yml 的内容为。

```
http_port: 80
```

❸ 定义和使用复杂变量

有时候我们需要使用的变量的值不是简单的字符串或者数字，而是一个对象。这时候定义的语法如下，格式为 YAML 的字典格式。

```
foo:
  field1: one
  field2: two
```

访问复杂变量中的子属性，可以利用中括号或者点号。

```
foo['field1']
foo.field1
```

❹ YAML 的陷阱

YAML 的陷阱是指某些时候 YAML 和 Ansible Playbook 的变量语法不能在一起好好工作了。这里仅发生在当冒号后面的值不以{开头的时候，如果有必要以{开头，则必须加上引号。总之，在 YAML 值的定义中，只要提示 YMAL 语法错误，就可以尝试加入引号来解决。

下面的代码会报错。

```
- hosts: app_servers
  vars:
     app_path: {{ base_path }}/22
```

解决办法是在"{"开始的值上加上引号。

```
- hosts: app_servers
  vars:
     app_path: "{{ base_path }}/22"
```

2. 远程主机的系统变量（Facts）

Ansible 会通过模块"setup"来搜集主机的系统信息，这些搜集到的系统信息叫作 Facts。

每个 Playbook 在执行前都会默认执行 setup 模块,所以这些 Facts 信息可以直接以变量的形式使用。

哪些 Facts 变量可以引用呢?可以通过在命令行上调用 setup 模块命令查看。

```
$ ansible all -m setup -u root
```

怎样在 Playbook 中使用 Facts 变量呢,答案是直接使用。

```
---
- hosts: all
  user: root
  tasks:
  - name: echo system
    shell: echo {{ ansible_os_family }}
  - name install ntp on Debian linux
    apt: name=git state=installed
    when: ansible_os_family == "Debian"
  - name install ntp on redhat linux
    yum: name=git state=present
    when: ansible_os_family == "RedHat"
```

❶ 使用复杂 Facts 变量

一般在系统中搜集到如下信息时,复杂的、多层级的 Facts 变量如何使用呢?

```
...
        "ansible_ens3": {
            "active": true,
            "device": "ens3",
            "ipv4": {
                "address": "10.66.192.234",
                "netmask": "255.255.254.0",
                "network": "10.66.192.0"
            },
            "ipv6": [
                {
                    "address": "2620:52:0:42c0:5054:ff:fef2:e2a3",
                    "prefix": "64",
                    "scope": "global"
                },
```

```
                {
                    "address": "fe80::5054:ff:fef2:e2a3",
                    "prefix": "64",
                    "scope": "link"
                }
            ],
            "macaddress": "52:54:00:f2:e2:a3",
            "module": "8139cp",
            "mtu": 1500,
            "promisc": false,
            "type": "ether"
        },
...
```

答案是可以通过下面的两种方式访问复杂变量中的子属性。

◎ 中括号。

```
{{ ansible_ens3["ipv4"]["address"] }}
```

◎ 点号。

```
{{ ansible_ens3.ipv4.address }}
```

❷ 关闭 Facts

搜集 Facts 信息会消耗额外时间，如果不需要 Facts 信息，则可以在 Playbook 中，通过关键字 gather_facts 来控制是否搜集远程系统的信息。如果不搜集系统信息，那么上面的 Facts 变量就不能在该 Playbook 中使用了。

```
- hosts: whatever
  gather_facts: no
```

3. 文件模板中使用的变量

template 模块在 Ansible 中十分常用，而它在使用的时候又没有显示地指定 template 文件中的值，所以有时候用户会对 template 文件中使用的变量感到困惑，所以在这里再重新强调下它的变量的使用。

❶ template 中变量的定义

在 Playbook 中定义的变量，可以直接在 template 中使用，同时 Facts 变量也可以直接在

template 中使用，当然也包含在 Inventory 里面定义的 Host 和 Group 变量。所有在 Playbook 中可以访问的变量，都可以在 template 文件中使用。

下面的 Playbook 脚本中使用了 template 模块来复制文件 index.html.j2，并且替换 index.html.j2 中的变量为 Playbook 中定义的变量值。

```yaml
---
- hosts: web
  vars:
    http_port: 80
    defined_name: "Hello My name is Jingjng"
  remote_user: root
  tasks:
  - name: ensure apache is at the latest version
    yum: pkg=httpd state=latest

  - name: Write the configuration file
    template: src=templates/httpd.conf.j2 dest=/etc/httpd/conf/httpd.conf
    notify:
    - restart apache

  - name: Write the default index.html file
    template: src=templates/index2.html.j2 dest=/var/www/html/index.html

  - name: ensure apache is running
    service: name=httpd state=started


  handlers:
    - name: restart apache
      service: name=httpd state=restarted
```

❷ template 中变量的使用

Ansible 模板文件使用变量的语法是 Python 的 template 语言 Jinja2。你不需要对 Jinja2 语言的语法有太多的了解，只需要知道 `{{ }}` 是用来引用变量的就可以了。

在下面的例子 template index.html.j2 中，直接使用了以下变量。

◎ 系统变量。

   {{ ansible_hostname }}

   {{ ansible_default_ipv4.address }}

◎ 用户自定义的变量。

   {{ defined_name }}

index.html.j2 文件的内容如下。

```
<html>
<title>Demo</title>
<body>
<div class="block" style="height: 99%;">
    <div class="centered">
        <h1>#46 Demo {{ defined_name }}</h1>
        <p>Served by {{ ansible_hostname }} ({{ ansible_default_ipv4.address }}).</p>
    </div>
</div>
</body>
</html>
```

4. 把运行结果当作变量使用——注册变量

把任务的执行结果当作一个变量的值也是可以的。这个时候就需要用到"注册变量",即把执行结果注册到一个变量中,待后面的任务使用。把执行结果注册到变量中的关键字是register,使用方法如下。

```
---

- hosts: web

  tasks:

    - shell: ls
      register: result
      ignore_errors: True

    - shell: echo "{{ result.stdout }}"
```

```
    when: result.rc == 5

- debug: msg="{{ result.stdout }}"
```

注册变量经常和 debug 模块一起使用，这样可以得到更多的关于执行错误的信息，以帮助用户调试。

5. 用命令行传递参数

为了使 Playbook 更灵活、通用性更强，允许用户在执行的时候传入变量的值，这时候就需要用到"额外变量"。

❶ 定义命令行变量

在 release.yml 文件中，hosts 和 user 都定义为变量，需要从命令行传递变量值。如果在命令行中不传入值，那么执行 Playbook 时会报错。

```
---
- hosts: '{{ hosts }}'
  remote_user: '{{ user }}'

  tasks:
    - ...
```

当然也可以直接在 Playbook 中定义普通的变量。像下面例子中的 test_name，如果在命令行中传入新的值，那么会覆盖 Playbook 中的值，未在命令行中的传入值也不会报错。

```
- hosts: localhost
  remote_user: root

  vars:
    test_name: "Value in playbook file"

  tasks:
    - debug: msg="{{ test_name }}"
```

❷ 使用命令行变量

在命令行里面传值的方法。

```
ansible-playbook e33_var_in_command.yml --extra-vars "hosts=web user=root"
```

还可以用 JSON 格式传递参数。

```
ansible-playbook e33_var_in_command.yml --extra-vars "{'hosts':'vm-rhel7-1', 'user':'root'}"
```

还可以将参数放在文件里面。

```
ansible-playbook e33_var_in_command.yml --extra-vars "@vars.json"
```

3.3.5　Playbook 也有逻辑控制语句

- ◎ when：条件判断语句，类似于编程语言中的 if。
- ◎ loop：循环语句，类似于编程语言的中的 while。
- ◎ block：把几个任务组成一个代码块，以便于针对一组操作的异常进行处理等操作。

1. 条件判断语句 when

when 是条件判断语句，类似于编程语言中的 if。

❶ When 的基本用法

有时候用户很可能需满足特定条件才执行某一个特定的步骤，例如在某一个特定版本的系统上安装软件包，或者只在磁盘空间不足的文件系统上执行清理操作一样。这些操作在 Playbook 中用 when 语句实现。

远程主机为 Debian Linux 对立刻关机。

```
tasks:
  - name: "shutdown Debian flavored systems"
    command: /sbin/shutdown -t now
    when: ansible_os_family == "Debian"
```

根据 Action 的执行结果，来决定接下来执行的任务。

```
tasks:
  - command: /bin/false
    register: result
    ignore_errors: True
  - command: /bin/something
```

```
    when: result|failed
  - command: /bin/something_else
    when: result|success
  - command: /bin/still/something_else
    when: result|skipped
```

远程中的系统变量 Facts 作为 when 的条件，用"|int"还可以转换返回值的类型。

```
---
- hosts: web
  tasks:
    - debug: msg="only on Red Hat 7, derivatives, and later"
      when: ansible_os_family == "RedHat" and ansible_lsb.major_release|int >= 6
```

❷ 条件表达式

假设 Playbook 中定义了变量 epic，那么如何在 when 中使用该变量写出各种条件表达式呢？我们用一些例子来说明。

定义变量 epic 如下。

```
vars:
  epic: true
```

基本款的 when 条件表达式。

```
tasks:
    - shell: echo "This certainly is epic!"
      when: epic
```

否定款。

```
tasks:
    - shell: echo "This certainly isn't epic!"
      when: not epic
```

变量定义款。

```
tasks:
    - shell: echo "I've got '{{ foo }}' and am not afraid to use it!"
      when: foo is defined
```

```
    - fail: msg="Bailing out. this play requires 'bar'"
      when: bar is not defined
```

数值表达款。

```
tasks:
   - command: echo {{ item }}
     with_items: [ 0, 2, 4, 6, 8, 10 ]
     when: item > 5
```

❸ 与 Include 一起用

```
- include: tasks/sometasks.yml
  when: "'reticulating splines' in output"
```

❹ 与 Role 一起用

```
- hosts: webservers
  roles:
     - { role: debian_stock_config, when: ansible_os_family == 'Debian' }
```

2. Loop 循环

❶ 标准循环

为了保持简洁，重复的任务可以用下面简写的方式。

```
- name: add several users
  user: name={{ item }} state=present groups=wheel
  with_items:
     - testuser1
     - testuser2
```

如果在变量文件中或者"vars"区域定义了一组列表变量 somelist，那么也可以这样做。

```
vars:
  somelist: ["testuser1", "testuser2"]
tasks:
 -name: add several user
  user: name={{ item }} state=present groups=wheel
  with_items: "{{somelist}}"
```

"with_items"用于迭代的 list 类型变量，不仅支持简单的字符串列表，如果你有一个哈希列表，那么可以用以下方式来引用子项。

```
- name: add several users
  user: name={{ item.name }} state=present groups={{ item.groups }}
  with_items:
    - { name: 'testuser1', groups: 'wheel' }
    - { name: 'testuser2', groups: 'root' }
```

注意：如果同时使用 when 和 with_items（或其他循环声明），那么 when 声明会为每个条目单独判断一次。

❷ 嵌套循环

循环也可以嵌套，用[]访问内层和外层的循环。

```
- name: give users access to multiple databases
  mysql_user: name={{ item[0] }} priv={{ item[1] }}.*:ALL append_privs=yes password=foo
  with_nested:
    - [ 'alice', 'bob' ]
    - [ 'clientdb', 'employeedb', 'providerd']
```

或者用点号（.）访问内层和外层的变量。

```
- name: give users access to multiple databases
  mysql_user: name={{ item.0 }} priv={{ item.1 }}.*:ALL append_privs=yes password=foo
  with_nested:
    - [ 'alice', 'bob' ]
    - [ 'clientdb', 'employeedb', 'providerd']
```

❸ 对哈希表使用循环

```
---
vars:
  alice:
    name: Alice Appleworth
    telephone: 123-456-7890
  bob:
    name: Bob Bananarama
    telephone: 987-654-3210
tasks:
  - name: Print phone records
```

```
    debug: msg="User {{ item.key }} is {{ item.value.name }}
({{ item.value.telephone }})"
      with_dict: "{{users}}"
```

❹ 对文件列表使用循环

with_fileglob 可以以非递归的方式来模拟匹配单个目录中的文件，如下所示。

```
tasks:

    # first ensure our target directory exists
    - file: dest=/etc/fooapp state=directory

    # copy each file over that matches the given pattern
    - copy: src={{ item }} dest=/etc/fooapp/ owner=root mode=600
      with_fileglob:
        - /playbooks/files/fooapp/*
```

3. Block 块

多个 action 组装成块后，可以根据不同条件执行一段语句。

```
tasks:
   - block:
       - yum: name={{ item }} state=installed
         with_items:
            - httpd
            - memcached

       - template: src=templates/src.j2 dest=/etc/foo.conf

       - service: name=bar state=started enabled=True

     when: ansible_distribution == 'CentOS'
     become: true
     become_user: root
```

组装成块后，处理异常会更加方便。

```
tasks:
  - block:
```

```yaml
    - debug: msg='i execute normally'
    - command: /bin/false
    - debug: msg='i never execute, cause ERROR!'
  rescue:
    - debug: msg='I caught an error'
    - command: /bin/false
    - debug: msg='I also never execute :-('
  always:
    - debug: msg="this always executes"
```

3.3.6 重用 Playbook

不能站在巨人肩膀上的编程语言不是好语言，支持重用机制可以节省在重复的工作上浪费的大量时间，同时还能提高可维护性。

Playbook 支持两种重用机制，一种是重用静态单个 Playbook 脚本，另外一种是重用实现特定功能的文件夹，类似于 Python 等编程语言中的包（Package）。

◎ include 语句：重用单个 Playbook 脚本，使用起来简单、直接。
◎ role 语句：重用实现特定功能的 Playbook 文件夹，使用方法稍复杂、功能强大。Ansible 还为 role 创建了一个共享平台 Ansible Galaxy，role 是 Ansible 最为推荐的重用和分享 Playbook 的方式。

1. include 语句

include 语句是基本的代码重用机制，主要重用任务，同时，include 还可将任务分割成多个文件，避免 Playbook 过于臃肿，使用户更关注于整体的架构，而不是实现的细节上。

❶ 基本的 include 语句用法

与其他语言的 include 语句一样，直接 include 即可。

```yaml
---
# possibly saved as tasks/firewall_httpd_default.yml

  - name: insert firewalld rule for httpd
    firewalld: port=80/tcp permanent=true state=enabled immediate=yes
```

main.yml 文件中调用 include 的方法。

```
tasks:
  - include: tasks/firewall_httpd_default.yml
```

❷ 在 include 语句中使用参数

这里有两个知识点：一是如何在被 include 的文件中定义参数，二是如何向 include 文件中传入参数。下面分别介绍。

◎ 在被 include 的文件中定义参数。在被 include 的文件 tasks/firewall_httpd_default.yml 中，使用 `{{ port }}` 定义了一个名字为 port 的参数：

```
---
- name: insert firewalld rule for httpd
  firewalld: port={{ port }}/tcp permanent=true state=enabled immediate=yes
```

◎ 向 include 文件中传入参数。

— 向执行的 Playbook 中传参数，可以加在行尾，使用空格分隔。

```
tasks:
 - include: tasks/firewall.yml port=80
 - include: tasks/firewall.yml port=3260
 - include: tasks/firewall.yml port=423
```

— 使用 YAML 字典传参数。

```
tasks:
- include: wordpress.yml
  vars:
    wp_user: timmy
    ssh_keys:
      - keys/one.txt
      - keys/two.txt
```

◎ 把一条任务简写成一个类 JSON 的形式传参数。

```
tasks:
 - { include: wordpress.yml, wp_user: timmy, ssh_keys: [ 'keys/one.txt', 'keys/two.txt' ] }
```

◎ 当然，在 Playbook 中已经定义了的参数，就不需要再显示传入值了，可以直接写成下面的形式。

```
---
- hosts: lb
```

```
  vars:
    port: 3206
  remote_user: root
  tasks:
    - include: tasks/firewall.yml
```

❹ include 语句相关的一些需要注意的地方

◎ 在 handler 里面加 include。在 handler 中加入 include 语句的语法如下。

```
handlers:
  - include: handlers/handlers.yml
```

注意：Ansible 1.9 及之前的版本是不能调用 include 里面的 handler 的，不过基于 Ansible 2.0+ 则可以调用 include 里面的 handler。所以在使用的时候要注意所安装的 Ansible 版本。

在下面的例子中，Ansible 1.9 不能调用 include 文件中的 handler "restart apache in include handlers"，而 Ansible 2.0+ 则可以调用该 hanlder。

```
- hosts: lb
  user: root
  gather_facts: no
  vars:
      random_number: "{{ 10000 | random }}"
  tasks:
  - name: Copy the /etc/hosts to /tmp/hosts.{{ random_number }}
    copy: src=/etc/hosts dest=/tmp/hosts.{{ random_number }}
    notify:
      - restart apache
      - restart apache in include handlers


  handlers:
    - include: handlers/handlers.yml
    - name: restart apache
      debug: msg="This is the handler restart apache"
```

◎ Ansible 允许全局（或者叫 Plays）加 include。然而这种使用方式并不推荐，因为它不支持嵌入 include，而且很多 Playbook 的参数也无法使用。

```
- name: this is a play at the top level of a file
  hosts: all
```

```
    remote_user: root

    tasks:

    - name: say hi
      tags: foo
      shell: echo "hi..."
# 全局 include，或者叫 Playbook include
- include: load_balancers.yml
- include: webservers.yml
- include: dbservers.yml
```

◎ 越来越强大而不稳定的 include。

为了使 include 功能更加强大，在每个新出的 Ansible 中都会添加一些新的功能。例如，在 2.0 中添加了 include 动态名字的 YAML，然而这样的用法有很多的限制，不够成熟，可能在更新版本的 Ansible 中又被去掉了，学习和维护成本很高。所以在需要使用更灵活的重用机制时，建议用下面介绍的 role。

2. role - Playbook 的 "Package"

role 有比 include 更强大灵活的代码重用和分享机制。include 类似于编程语言中的 include，是重用单个文件的，重用的功能有限。

而 role 类似于编程语言中的 "Package"，可以重用一组文件，形成完整的功能。例如，安装和配置 Apache，既需要用任务实现安装包和复制模板，也需要 httpd.conf 和 index.html 的模板文件，以及 handler 文件实现重启功能。这些文件都可以放在一个 role 里面，以供不同的 Playbook 文件重用。

Ansible 十分提倡在 Playbook 中使用 role，并且提供了一个分享 role 的平台 Ansible Galaxy（https://galaxy.ansible.com/）。在 Galaxy 上可以找到别人写好的 role。在后面的章节中，我们将详细介绍如何使用它。

❶ 定义 role 完整的目录结构

在 **Ansible** 中，通过遵循特定的目录结构，就可以实现对 **role** 的定义。具体遵循的目录结构是什么呢？看下面的例子。

下面的目录结构定义了一个 role：名字为 myrole。Playbook 文件中的 site.yml，调用了这个 role。

Ansible 并不要求 role 包含上述所有的目录及文件，可以根据 role 的功能，加入对应的目录和文件即可。下面是每个目录和文件的功能：

◎ 如果 roles 文件 role/x/tasks/main.yml 存在，则文件中列出的任务都将被添加到 Play 中。
◎ 如果文件 roles/x/handlers/main.yml 存在，则文件中列出 handler 都将被添加到 Play 中。
◎ 如果文件 role/x/vars/main.yml 存在，则文件中列出的变量都将被添加到 Play 中。
◎ 如果文件 role/x/defaults/ain.yml 存在，则文件中列出的变量都会被添加到 Play 中。
◎ 如果文件 role/x/meta/main.yml 存在，则文件中列出的 所有依赖的 role 都将被添加到 Play 中。
◎ 此外，下面的文件不需要绝对或者是相对路径，和放在同一个目录下的文件一样，直接使用即可。
　　— copy 或者 script 使用 roles/x/files/ 下的文件。
　　— template 使用 roles/x/templates 下的文件。
　　— include 使用 roles/x/tasks 下的文件。

而在写 role 的时候，一般都要包含 role 入口文件 roles/x/tasks/main.yml，其他的文件和目录，可以根据需求选择是否加入。

学会写一个完整功能的 role 是一个相对复杂的过程，也是 Ansible 中稍高级的使用方法。在本节，我们重点介绍如何使用别人已经写好的 role。在后面的章节中，我们会通过具体的示例来逐步介绍写 role 所需的知识。

❷ 带参数的 role

参数在 role 中是如何定义的呢？

下面定义一个带参数的 role，名字是 myrole，其目录结构如下。

```
main.yml
 roles
  myrole
    tasks
      main.yml
```

在 roles/myrole/tasks/main.yml 中，使用 **{{ }}** 定义的变量就可以了。

```
---
- name: use param
  debug: msg="{{ param }}"
```

❸ 使用带参数的 role

在 main.yml 中可以用如下方法使用 myrole 了。

```
---

- hosts: webservers
  roles:
    - { role: myrole, param: 'Call some_role for the 1st time' }
    - { role: myrole, param: 'Call some_role for the 2nd time' }
```

或者写成 YAML 字典格式：

```
---

- hosts: webservers
  roles:
    - role: myrole
```

```
      param: 'Call some_role for the 1st time'
    - role: myrole
      param: 'Call some_role for the 2nd time'
```

❹ role 指定默认的参数

指定默认参数后，如果在调用时传参数了，那么就使用传入的参数值；如果在调用的时候没有传参数，那么就使用默认的参数值。

指定默认参数很简单，以上面的 myrole 为例。

```
main.yml
roles:
  myrole
    tasks
      main.yml
    defaults
      main.yml
```

在 roles/myrole/defaults/main.yml 中，使用 YAML 字典定义语法定义的 param 的值如下。

```
param: "I am the default value"
```

这样在 main.yml 中，下面两种调用方法都可以。

```
---

- hosts: webservers
  roles:
    - myrole
    - { role: myrole, param: 'I am the value from external' }
```

在本节，只需了解默认参数的作用即可，更多关于 role 的写法请参考本书第 6 章。

❺ role 与条件语句 when 一起

下面的例子中，myrole 只在 Red Hat 系列的主机上才能执行。

```
---

- hosts: webservers
  roles:
    - { role: myrole, when: "ansible_os_family == 'RedHat'" }
```

同样也可以写成 YAML 字典格式。

```yaml
---

- hosts: webservers
  roles:
    - role: my_role
      when: "ansible_os_family == 'RedHat'"
```

❻ role 和任务的执行顺序

如果一个 Playbook 中同时出现 role 和任务，那么它们的调用顺序是怎样的呢？

先揭晓答案，再根据实例来验证：

pre_tasks > role > tasks > post_tasks。

```yaml
---

- hosts: lb
  user: root

  pre_tasks:
    - name: pre
      shell: echo 'hello'

  roles:
    - { role: some_role }

  tasks:
    - name: task
      shell: echo 'still busy'

  post_tasks:
    - name: post
      shell: echo 'goodbye'
```

执行的结果如下。

```
PLAY [lb] *************************************************************
*********
```

```
    TASK [setup] **************************************************
*******
    ok: [rhel7u3]

    TASK [pre] ****************************************************
********
    changed: [rhel7u3]

    TASK [some_role : some role] **********************************
*****
    ok: [rhel7u3] => {
        "msg": "Im some role"
    }

    TASK [task] ***************************************************
********
    changed: [rhel7u3]

    TASK [post] ***************************************************
********
    changed: [rhel7u3]

    PLAY RECAP ****************************************************
********
    rhel7u3                    : ok=5    changed=3    unreachable=0    failed=0
```

3.3.7 用标签，实现执行 Playbook 中的部分任务

如果 Playbook 文件比较大，并且在执行的时候只是想执行部分功能，那么这个时候有没有解决方案呢？Playbook 提供了标签（tags）可以实现部分运行。

1. 标签的基本用法

例如，文件 example.yml 标记了两个标签：packages 和 configuration。

```yaml
tasks:

  - yum: name={{ item }} state=installed
    with_items:
       - httpd
    tags:
       - packages

  - name: copy httpd.conf
    template: src=templates/httpd.conf.j2 dest=/etc/httpd/conf/httpd.conf
    tags:
       - configuration

  - name: copy index.html
    template: src=templates/index.html.j2 dest=/var/www/html/index.html
    tags:
       - configuration
```

我们在执行的时候，如果不加任何 tag 参数，那么会执行所有标签对应的任务。

```
ansible-playbook example.yml
```

如果指定执行安装部分的任务，则可以利用关键字 tags 指定需要执行的标签名。

```
ansible-playbook example.yml --tags "packages"
```

如果指定不执行 tag packages 对应的任务，则可以利用关键字 skip-tags。

```
ansible-playbook example.yml --skip-tags "configuration"
```

2. 特殊的标签

❶ always

标签的名字是用户自定义的，但是如果把标签的名字定义为 always，那么就有点特别了。只要在执行 Playbook 时，如果没有明确指定不执行 always 标签，那么 always 标签所对应的任务就始终会被执行。

在下面的例子中，即便只指定执行 packages 标签，always 标签也会被执行。

```yaml
tasks:
```

```yaml
    - debug: msg="Always print this debug message"
      tags:
        - always

    - yum: name= state=installed
      with_items:
        - httpd
      tags:
        - packages

    - template: src=templates/httpd.conf.j2 dest=/etc/httpd/conf/httpd.conf
      tags:
        - configuration
```

只指定运行 **packages** 标签时，还是会执行 **always** 标签对应的任务。

```
ansible-playbook tags_always.yml --tags "packages"
```

❷ "tagged"、"untagged" 和 "all"

```yaml
tasks:

  - debug: msg="I am not tagged"
    tags:
      - tag1

  - debug: msg="I am not tagged"
```

分别指定 --tags 为 "tagged"、"untagged" 和 "all" 试下效果吧。

下面的例子中，利用 "--tags tagged" 来执行所有标记了标签的任务，无论标记的标签的名字是什么。

```
ansible-playbook tags_tagged_untagged_all.yml --tags tagged
```

下面的例子中，利用 "--tags untagged" 来执行所有没有标记标签的任务，无论标记的标签的名字是什么：

```
ansible-playbook tags_tagged_untagged_all.yml --tags untagged
```

下面的例子中，利用 "--tags all" 来执行所有任务。

```
ansible-playbook tags_tagged_untagged_all.yml --tags all
```

3. 在 include 和 role 中分别使用标签

include 语句指定执行的标签的语法：

```
- include: foo.yml
  tags: [web,foo]
```

调用 role 中的标签的语法如下。

```
roles:
  - { role: webserver, port: 5000, tags: [ 'web', 'foo' ] }
```

3.4 更多的 Ansible 模块

- 介绍两类模块：Core 模块和 Extra 模块。
- Extra 模块的配置和使用方法。
- 通过命令行查看模块的用法。

3.4.1 模块的分类

在 Ansible 模块文档上查看单个模块的时候，每一个模块文档的底部都会标识，这是"Core Module"，还是"Extra Module"。

比如，yum 就是一个 Core 模块，而 archive 就是一个 Extra 模块。

1. Core 模块（核心模块）

- 不需要额外下载和配置，安装 Ansible 后就可以直接使用的。
- 比较常用的模块。
- 经过严格测试的模块。

2. Extra 模块（额外模块）

- 需进行下载和额外的配置才能使用。
- 次常用的模块。
- 还有可能存在 bug 的模块。

3. Core 模块和 Extra 模块的变迁

Ansible 团队一直致力于把成熟的长期使用没有问题的模块放入 Core 模块中，方便用户的使用。所以当你的 Playbook 运行时，如果报错但没有相应的模块，那么你只要知道问题可能出现在使用了 Extra 模块而没有进行相关的配置就可以了。下面给出配置 Extra 模块的解决方法。

3.4.2　Extra 模块的使用方法

使用 Exra 模块前需要进行下面的配置，以便可以在命令行或者是 Playbook 中使用。配置后的 Extra 模块的使用方法和 Core 模块的使用方法相同。

❶ 下载 Ansible Module Extra 项目。

```
git clone https://github.com/ansible/ansible-modules-extras.git
```

这里下载到了/home/jshi/software/目录下，后面会用到这个目录。

❷ 修改配置文件或者环境变量。

方法 1：修改 Ansible 默认配置文件/etc/ansible/ansible.cfg。

修改 Ansible 配置文件/etc/ansible/ansible.cfg，添加下面一行。

```
library    = /home/jshi/software/ansible-modules-extras/
```

方法 2：修改 Ansible 当前目录下配置文件 ansible.cfg。

修改 Ansible Playbook 当前目录下的配置文件 ansible.cfg，使其只对当前目录的 Playbook 生效。对所有其他目录，包括父目录和子目录的 Playbook 都不生效。

```
library/ansible-modules-extras
ansible.cfg
use_extra_module.yml
subfolder/use_extra_module_will_throw_error.yml
```

在当前目录的 ansible.cfg 中，可以使用相对路径。

```
library = library/ansible-modules-extras/
```

方法 3：改环境变量。

```
export ANSIBLE_LIBRARY=/project/demo/demoansible/library/ansible-module-extras
```

如果需要环境变量在重启后生效,那么必须在~/.bashrc 中声明 ANSIBLE_LIBRARY 变量。

```
$ echo >>~/.bashrc <<EOF

export ANSIBLE_LIBRARY=/project/demo/demoansible/library/ansible-module-extras

EOF

$ source ~/.bashrc
```

3.4.3 命令行查看模块的用法

类似 bash 命令的 man,Ansible 也可以通过命令行查看模块的用法。命令是 ansible-doc,语法如下。

```
ansible-doc module_name
```

Core 模块可以在任何目录下执行,例如查看 yum 的用法:

```
ansible-doc yum
```

Extra 模块必须在配置了 Extra 模块的目录下查看其用法。

```
ansible-doc archive
```

3.5 最佳使用方法

3.5.1 写 Playbook 的原则

Ansible 为了降低 Playbook 的维护成本,提高 Playbook 的可维护性,提倡以下两个原则:

◎ 鼓励文件的重用,尽量使用 include 和 role 避免重复的代码。
◎ 尽量把大的文件分成小的文件。

3.5.2 参考别人的 Playbook

能够学会快速使用别人的成果并高效地分享自己的成果，才是好码农。在你动手从头开始写一个 Playbook 之前，不如先参考一下别人的成果吧。

1. 官方例子

Ansible 官方提供了一些比较常用的、经过测试的 Playbook 例子：

https://github.com/ansible/ansible-examples

2. Playbook 分享平台 Ansible Galaxy

此外，Ansible 还提供了一个 Playbook 的分享平台，平台上的例子是 Ansible 用户自己上传的，在没有思路的时候可以参考，不过一定要再重新严谨的测试：

https://galaxy.ansible.com/

第 4 章

Ansible Playbook 杂谈

本章重点

4.1 再谈 Ansible 变量
4.2 使用 lookup 访问外部文件或数据库中的数据
4.3 过滤器
4.4 测试变量或表达式是否符合条件
4.5 认识插件

4.1 再谈 Ansible 变量

在前面的章节中已经介绍过很多变量，为什么在这里还要重复这个话题呢，因为变量实在太重要了。本章将对所有的 Ansible 支持的变量，根据其优先级，依次介绍其用法和作用域。

4.1.1 变量的作用域

- Global，作用域为全局：
  - Ansible 配置文件中定义的变量。
  - 环境变量。
  - ansible/ansible-playbook 命令行中传进来的变量。
- Play，作用域为 Play（一个 Playbook 由多个 Play 构成）：
  - Play 中 vars 关键字下定义的变量。
  - 通过模块 include_vars 定义的变量。
  - role 在文件 default/main.yml 和 vars/main.yml 中定义的变量。
- Host，作用域为某个主机：
  - 定义在主机清单中的变量。
  - 主机的系统变量。
  - 注册变量。

4.1.2 变量的优先级

表 4.1 变量的优先级

Ansible 变量的优先级（由低到高）
role defaults
dynamic inventory variables
inventory variables
inventory group_vars

续表

Ansible 变量的优先级（由低到高）
inventory host_vars
playbook group_vars
playbook host_vars
host facts
registered variables
set_facts
play variables
play vars_prompt
play vars_files
role variables and include variables
block variables
task variables
extra variables

从上面的变量优先级表中，我们可以总结出大体的规律，除了 role defaults 变量外，其他变量的作用域越小越精确，变量的优先级越高。inventory 中的全局变量可以被 Play 中的变量覆盖，Playbook 中的变量可以被 host 变量覆盖，host 变量又可以被 task 变量覆盖。变量优先级最高的是 extra 变量，又叫命令行变量，只在某一次执行时生效的变量优先级最高。

变量的优先级只有在变量重名的时候才需要区分，所以建议除了 role defaults 变量外，尽量不要有相同名字的变量。

上面的绝大部分变量，都已经在前面的章节简单介绍过，或者用过，在本章，我们会再次介绍所有变量的用法，以便读者加深印象，区分各种变量的用法。

1. role defaults

role x 的默认变量放在文件 roles/x/defaults/main.yml 中。

```
---
# file: roles/x/defaults/main.yml
# if not overridden in inventory or as a parameter, this is the value that will be used
  http_port: 80
```

2. inventory vars

在 inventory 文件中定义的变量。

```
#file: /etc/ansible/hosts
host1 ansible_port=5555 ntp_server=inventory.ntp.com

[all:vars]
ntp_server=inventory_vars.ntp.com
```

3. inventory group_vars

有两个地方可以定义 group_vars：一是在 inventory 文件中直接定义；二是在 inventory 文件同级的子文件夹 group_vars 下定义，放在 group 同名的文件中。

在 Ansible 中，默认的 inventory 文件为/etc/ansible/hosts。

```
# file: /etc/ansible/hosts
[group1:vars]
ntp_server=inventory_host_vars.ntp.com
```

所有 group 中都生效的变量放在文件/etc/ansible/group_vars/all 中。

```
---
# file: /etc/ansible/group_vars/all
# this is the site wide default
ntp_server: default-time.example.com
```

group1 的变量放在文件/etc/ansible/group_vars/group1 中。

```
---
# file: /etc/ansible/group_vars/group1
ntp_server: boston-time.example.com
```

group1 中的主机在执行时，采用的是 group1 中的 ntp_server 变量。因为该变量作用域更精确，所以优先级会更高。

4. inventory host_vars

有两个地方可以定义 host_vars：一是在 inventory 文件中直接定义；二是在 inventory 文件同级的子文件夹 host_vars 下，与 host 同名的文件中定义。

在 Ansible 中，默认的 inventory 文件为/etc/ansible/hosts。

```
# file: /etc/ansible/hosts
host1 ntp_server=inventory_host_vars.ntp.com
```

可以放在 inventory 文件同级的子目录 host_vars 下，与 Host 同名的文件中。

```
---
# file: /etc/ansible/host_vars/host1
ntp_server: override.example.com
```

5. playbook group_vars

和 Playbook 文件同级的子目录下定义的变量。例如，当前的 Playbook 放在~/playbooks 目录下，那么 Group "group1" 的变量放在下面的文件中：

```
---
# file: ~/playbooks/group_vars/group1
ntp_server: override.example.com
```

6. playbook host_vars

Playbook 文件同级的子目录 Host-vars 下定义的变量。例如，当前的 Playbook 放在~/playbooks 目录下，那么对应的 Host "host1" 的变量放在下面的文件中：

```
---
# file: ~/playbooks/host_vars/host1
ntp_server: override.example.com
```

7. host facts

Ansible 在执行 Playbook 时，都会搜集远程主机的信息。这些关于主机的系统变量都可以在 Playbook 中直接使用。具体搜索了哪些变量，可以通过下面的命令得到：

```
ansible host1 -m setup
```

8. play vars

```
- hosts: web
  vars:
    http_port: 80
    defined_name: "Hello My name is Jingjng"
```

9. play vars_prompt

vars_prompt 是需要用户在执行 Playbook 的时候输入变量值的变量。

```
---
- hosts: all
  remote_user: root
  gather_facts: no

  vars:
    test: "camelot"

  vars_prompt:
    - name: "name"
      prompt: "what is your name?"
    - name: "favcolor"
      prompt: "what is your favorite color?"
  tasks:
    - debug: msg="Hello {{name}}, your favorite color is {{favcolor}}"
```

10. play vars_files

把变量单独放在一个文件中，通过关键字 var_files 从文件中加载进来的变量就是 Play var_files。

```
- hosts: web
  var_files:
    - apache_vars.yml
```

11. registered vars

将执行结果注册到一个动态值的变量中，这个变量就是 registered vars。

```
    - shell: ls
      register: result
      ignore_errors: True

    - debug: msg="{{ result.stdout }}"
```

12. set_facts

set_facts 是一个模块的名字，在任务中通过 set_facts 加入一些 Facts 变量。

```
- set_fact:
    one_fact: "something"
    other_fact: "{{ local_var }}"
```

13. role and include vars

role vars。

```
---
# file: roles/x/vars/main.yml
# this will absolutely be used in this role
http_port: 80
```

14. role include vars

在 role/x/tasks/main.yml include 中，通过关键字 include 加载进来的变量。

```
---
# file: roles/x/vars/apache.yml
# this will absolutely be used in this role
http_port: 80
```

```
---
# file: roles/x/tasks/main.yml
- name: Add apache variables
  include_vars: "apache.yml"
```

15. block vars

只能在 Playbook 的任务中的某个 block 里定义和使用的变量。

```
  tasks:
- block:
    - yum: name={{ service }} state=installed
    - service: name={{ status }} state=started enabled=True
  vars:
    service: httpd
```

```
    status: started
```

16. task vars

只能在该任务里面使用的变量。

```
tasks:
 - debug: msg="{{service}} is {{status}}"
   vars:
     service: httpd
     status: running
```

17. extra vars

通过命令行传进来的变量。

```
ansible... -e "ansible_ssh_user=<user>"
```

4.2 使用 lookup 访问外部文件或数据库中的数据

Ansible Playbook 允许用户使用自定义的变量，不过当变量过大，或者太复杂时，无论是在 Playbook 中通过 vars 定义，还是在单独的变量文件中定义，可读性都比较差，而且不够灵活。

有了 lookup 就能解决这类难题，lookup 既能够读取 Ansible 管理节点上文件系统的文件内容到 Ansible 变量中，也可以读取配置的数据库中的内容。

下面是 lookup 的基本使用方法，将 Ansible 管理节点上的文件 data/plain.txt 的内容读取出来，并赋值给变量 "contents"。"file" 告诉 lookup 读取对象的类型是 File，直接读取文件内容，无须做特别的处理。

```
---
- hosts: all
  remote_user: root
  gather_facts: false
  vars:
    contents: "{{ lookup('file', 'data/plain.txt') }}"
```

```
    tasks:

      - debug: msg="the value of data/plain.txt is {{ contents }}"
```

执行结果如下。

```
PLAY [all] **********************************************************
*******

TASK [debug] ********************************************************
********
ok: [jshi-test-02.rhev] => {
    "msg": "the value of data/plain.txt is Hello, You will my text."
}

PLAY RECAP **********************************************************
********
jshi-test-02.rhev          : ok=1    changed=0    unreachable=0    failed=0
```

其实 lookup 还很很多更强大的功能，下面一一列举。

4.2.1 lookup 读取文件

上面的例子 "{{ lookup('**file**', 'data/plain.txt') }}" 使用了 file 类型的 lookup，是最简单的 lookup 用法。

4.2.2 lookup 生成随机密码

第一次执行时，如果密码文件 /tmp/password/kitty 不存在，lookup 会生成长度为 5 的随机密码，存储在文件中。如果密码文件存在，那么直接读取该文件中的内容作为密码。这样在创建用户的时候生成密码就很方便了。

```
---
- hosts: all
  remote_user: root
  vars:
```

```
      password: "{{ lookup('password', '/tmp/password/kitty length=5') }}"

  tasks:
    - debug: var=password
```

执行结果如下。

```
[jshi@jjshi 10_lookup]$ ansible-playbook site_lookup_password.yml

PLAY [all] **********************************************************
********

TASK [debug] ********************************************************
*******
ok: [jshi-test-02.rhev] => {
    "password": "bsl82"
}

PLAY RECAP **********************************************************
********
jshi-test-02.rhev          : ok=1    changed=0    unreachable=0    failed=0
```

4.2.3　lookup 读取环境变量

env 类型的 lookup 可以读取 Linux 上的环境变量，如下面的示例。

```
---
- hosts: all
  remote_user: root

  tasks:
    - debug: msg="{{ lookup('env','HOME') }} is an environment variable"
```

执行结果如下。

```
TASK [debug] ********************************************************
*******
ok: [jshi-test-02.rhev] => {
    "msg": "/home/jshi is an environment variable"
}
```

4.2.4　lookup 读取 Linux 命令的执行结果

pipe 类型的 lookup 可以将 Linux 上命令的执行结果读取到 Ansible 中：

```
---
- hosts: all
  remote_user: root

  tasks:
    - debug: msg="{{ lookup('pipe','date') }} is the raw result of running this command"
```

执行的结果为。

```
TASK [debug] ************************************************************
ok: [jshi-test-02.rhev] => {
    "msg": "Sat Feb 11 21:00:58 CST 2017 is the raw result of running this command"
}
```

4.2.5　lookup 读取 template 变量替换后的文件

template 类型的 lookup 可以将一个 template 文件经过变量替换后的内容读取到 Ansible 中。当然，如果在 template 文件中有未定义的变量，则会报错。

```
---
- hosts: all
  remote_user: root
  vars:
    name: "Crystal"

  tasks:
    - debug: msg="{{ lookup('template', 'data/some_template.j2') }} is a value from evaluation of this template"
```

模板文件 data/som_templates.j2 的内容如下。

```
I am the template for test lookup template, and my name is {{name}}
```

执行的结果如下所示，lookup 可以得到模板文件中变量替换后的结果。

```
[jshi@jjshi 10_lookup]$ ansible-playbook site_lookup_template.yml

PLAY [all] ***************************************************************

TASK [setup] *************************************************************
ok: [jshi-test-02.rhev]

TASK [debug] *************************************************************
ok: [jshi-test-02.rhev] => {
    "msg": "I am the template for test lookup template, and my name is Crystal\n is a value from evaluation of this template"
}

PLAY RECAP ***************************************************************
jshi-test-02.rhev          : ok=2    changed=0    unreachable=0    failed=0
```

4.2.6 lookup 读取配置文件

lookup 支持读取两种类型的配置文件：ini 和 Java 的 properties。下面分别介绍这两种文件的读取方法。

ini 类型的 lookup 默认读取配置文件的类型是 ini。

假设有 ini 类型文件 data/users.ini，内容如下。

```
[production]
# My production information
user=robert
pass=somerandompassword


[integration]
# My integration information
```

```
user=gertrude
pass=anotherpassword
```

lookup 的使用方法如下。

```
---
- hosts: all
  remote_user: root

  tasks:
    - debug:
        msg: "User in integration is {{ lookup('ini', 'user section=integration file=data/users.ini') }}"
    - debug:
        msg: "User in production is {{ lookup('ini', 'user section=production file=data/users.ini') }}"
```

执行的结果如下。

```
TASK [debug] ****************************************************************
ok: [jshi-test-02.rhev] => {
    "msg": "User in integration is gertrude"
}

TASK [debug] ****************************************************************
ok: [jshi-test-02.rhev] => {
    "msg": "User in production is robert"
}
```

读取 properties 类型文件时，需要加一个额外的参数来告诉 lookup，这是 properties 类型的文件。

data/user.properties 中的文件内容如下。

```
user.name=robert
user.pass=somerandompassword
```

读取该 Properties 类型文件的方法如下。

```
---
- hosts: all
```

```
    remote_user: root

    tasks:
      - debug: msg="user.name is {{ lookup('ini', 'user.name type=properties
file=data/user.properties') }}"
```

执行的结果如下。

```
TASK [debug] *******************************************************
*******
    ok: [jshi-test-02.rhev] => {
      "msg": "user.name is robert"
    }
```

ini 类型的参数格式。

```
lookup('ini', 'key [type=<properties|ini>] [section=section] [file=file.ini]
[re=true] [default=<defaultvalue>]')
```

每个参数都有默认值，所以在使用 ini 类型的 lookup 时，每一个参数都是可选的，没有传入的参数会使用默认值。每个参数的含义和默认值如表 4.2 所示。

表 4.2 参数的含义和默认值

参数名	默认值	参数含义
type	ini	文件的类型
file	ansible.ini	加载文件的名字
section	global	默认的在哪个 section 里面查找 key
re	False	key 的正则表达式
default	empty string	key 不存在时的返回值

4.2.7　lookup 读取 CSV 文件的指定单元

```
---
- hosts: all
  remote_user: root

  tasks:
    - debug: msg="The atomic number of Lithium is {{ lookup('csvfile', 'Li
```

```
file=elements.csv delimiter=,') }}"
    - debug: msg="The atomic mass of Lithium is {{ lookup('csvfile', 'Li
file=elements.csv delimiter=, col=2') }}"
```

执行结果如下。

```
TASK [debug] ***************************************************************
*********
ok: [jshi-test-02.rhev] => {
    "msg": "The atomic number of Lithium is 3"
}

TASK [debug] ***************************************************************
*********
ok: [jshi-test-02.rhev] => {
    "msg": "The atomic mass of Lithium is 6.94"
}
```

csvfile 类型的 lookup 的语法如下。

```
Lookup('csvfile', 'key arg1=val1 arg2=val2 …')
```

key 是 column 0 中某一个单元的值，其他参数都是可选的。支持的参数如表 4.3 所示。

表 4.3 支持的参数

参数名	默认值	含义
file	ansible.csv	加载文件的名字
col	1	输出的列的索引，索引从 0 开始计数
delimiter	TAB	CSV 文件的分隔符；tab 可以用 TAB 或者 t 来表示
default	空字符串	如何元素不存在时的返回值
encoding	utf-8	CSV 文件的编码

4.2.8 lookup 读取 DNS 解析的值

dig 类型的 lookup 可以向 DNS 服务器查询指定域名的 DNS 记录。它可以查询任何 DNS 记录，包括正向查询和反向查询。

下面的文件使用 dig 类型的 lookup 查看 DNS，不仅使用了正向 DNS 解析查询，还使用了反向 DNS 解析查询，此外还可以指定查询的 DNS 服务器。

```
---
- hosts: all
  remote_user: root
  gather_facts: false

  tasks:
  #正向查询某一个域名的 DNS 记录
    - debug: msg="The IPv4 address for baidu.com. is {{ lookup('dig', 'baidu.com.') }}"
    - debug: msg="The TXT record for baidu.com. is {{ lookup('dig', 'baidu.com.', 'qtype=TXT') }}"
    - debug: msg="The TXT record for baidu.com. is {{ lookup('dig', 'baidu.com./TXT') }}"
    - debug: msg="One of the MX records for 163.com. is {{ item }}"
      with_items: "{{ lookup('dig', '163.com./MX', wantlist=True) }}"
```

上面的例子中，dig lookup 的执行结果如下。

```
TASK [debug] ****************************************************
********
ok: [jshi-test-02.rhev] => {
    "msg": "The IPv4 address for baidu.com. is 123.125.114.144,180.149.132.47,220.181.57.217,111.13.101.208"
}

TASK [debug] ****************************************************
********
ok: [jshi-test-02.rhev] => {
    "msg": "The TXT record for baidu.com. is google-site-verification=GHb98-6msqyx_qqjGl5eRatD3QTHyVB6-xQ3gJB5UwM,v=spf1 include:spf1.baidu.com include:spf2.baidu.com include:spf3.baidu.com a mx ptr -all"
}

TASK [debug] ****************************************************
********
ok: [jshi-test-02.rhev] => {
```

```
        "msg": "The TXT record for baidu.com. is v=spf1 include:spf1.baidu.com
include:spf2.baidu.com include:spf3.baidu.com a mx ptr -all,google-site-
verification=GHb98-6msqyx_qqjGl5eRatD3QTHyVB6-xQ3gJB5UwM"
    }

    TASK [debug] ****************************************************
*******
    ok: [jshi-test-02.rhev] => (item=10 163mx01.mxmail.netease.com.) => {
        "item": "10 163mx01.mxmail.netease.com.",
        "msg": "One of the MX records for 163.com. is 10 163mx01.mxmail.
netease.com."
    }
    ok: [jshi-test-02.rhev] => (item=50 163mx00.mxmail.netease.com.) => {
        "item": "50 163mx00.mxmail.netease.com.",
        "msg": "One of the MX records for 163.com. is 50 163mx00.mxmail.
netease.com."
    }
    ok: [jshi-test-02.rhev] => (item=10 163mx03.mxmail.netease.com.) => {
        "item": "10 163mx03.mxmail.netease.com.",
        "msg": "One of the MX records for 163.com. is 10 163mx03.mxmail.
netease.com."
    }
    ok: [jshi-test-02.rhev] => (item=10 163mx02.mxmail.netease.com.) => {
        "item": "10 163mx02.mxmail.netease.com.",
        "msg": "One of the MX records for 163.com. is 10 163mx02.mxmail.
netease.com."
    }
```

反向查询 DNS。

```
    #反向查询某一个 IP 对应的 DNS 记录
    - debug: msg="Reverse DNS for 23.214.122.249 is {{ lookup('dig',
'23.214.122.249/PTR') }}"
    - debug: msg="Reverse DNS for 23.214.122.249 is {{ lookup('dig',
'249.122.214.23.in-addr.arpa./PTR') }}"
    - debug: msg="Reverse DNS for 23.214.122.249 is {{ lookup('dig',
'249.122.214.23.in-addr.arpa.', 'qtype=PTR') }}"
```

反向查询的结果如下。

```
    TASK [debug] **********************************************************
********
ok: [jshi-test-02.rhev] => {
    "msg": "Reverse DNS for 23.214.122.249 is a23-214-122-249.deploy.
static.akamaitechnologies.com."
}

    TASK [debug] **********************************************************
*******
ok: [jshi-test-02.rhev] => {
    "msg": "Reverse DNS for 23.214.122.249 is a23-214-122-249.deploy.
static.akamaitechnologies.com."
}

    TASK [debug] **********************************************************
********
ok: [jshi-test-02.rhev] => {
    "msg": "Reverse DNS for 23.214.122.249 is a23-214-122-249.deploy.
static.akamaitechnologies.com."
}
```

指定查询的 DNS 服务器。

```
    # 指定查询的 DNS 服务器为 8.8.8.8
    - debug: msg="Querying 8.8.8.8 for IPv4 address for baidu.com. produces
{{ lookup('dig', 'baidu.com', '@8.8.8.8') }}"
```

执行的结果如下。

```
    TASK [debug] **********************************************************
*********
ok: [jshi-test-02.rhev] => {
    "msg": "Querying 8.8.8.8 for IPv4 address for baidu.com. produces
180.149.132.47,220.181.57.217,111.13.101.208, 123.125.114.144"
}
```

4.2.9　更多的 lookup 功能

lookup 不限于支持上述几种数据的读取和导入，这里只是列举了几个较为常用的 lookup 的功能。如果希望了解更多关于 lookup 的功能，请参考下面的 Ansible 官方文档中关于 lookup 的部分，需要注意的是，有些 lookup 的功能需要额外的 Python 包支持：

http://docs.ansible.com/ansible/playbooks_lookups.html#the-credstash-lookup

4.3　过滤器

过滤器（filter）是 Python 模板语言 Jinja2 提供的模块，可以用来操作数据。它在 Ansible 的管理节点上执行并操作数据，而不是在远程的目标主机上。

过滤器和 lookup 类似，都是在 {{ }} 中使用。不同类型的过滤器的功能差距很大。过滤器是 Ansible 使用的模板语言 Jinja2 的内置功能。在 Ansible 中，不仅可以使用 Jinja2 自带的过滤器，还可以使用 Ansible 提供的过滤器，以及用户根据自己的需要自定义的过滤器。

本节重点介绍 Ansible 提供的过滤器，Jinja2 自带的过滤器请参考下面的链接：

http://jinja.pocoo.org/docs/2.9/templates/#builtin-filters。

4.3.1　过滤器使用的位置

下面用一个最简单的过滤器来说明过滤器的语法，quote 过滤器的功能是给字符串加引号。

```
---
- hosts: localhost
  remote_user: root
  gather_facts: false

  vars:
    my_test_string: "This is the test string"

  tasks:
```

```yaml
- name: "quote {{ my_test_string }}"
  debug: msg="echo {{ my_test_string | quote }}"
```

输出的结果如下。

```
[jshi@jjshi 11_filter]$ ansible-playbook site_filter_quote.yml

PLAY [localhost] ****************************************************
********

TASK [quote This is the test string] ********************************
*******
ok: [localhost] => {
    "msg": "echo 'This is the test string'"
}

PLAY RECAP **********************************************************
********
localhost                  : ok=1    changed=0    unreachable=0    failed=0
```

4.3.2 过滤器对普通变量的操作

❶ default：为没有定义的变量提供默认值。

因为变量 some_undefined_varible 表示没有定义，所以下面的任务会输出 Default。

```
- debug: msg="{{ some_undefined_variable | default("Default") }}"
```

输出的结果如下。

```
TASK [debug] ********************************************************
*******
ok: [localhost] => {
    "msg": "Default"
}
```

❷ omit：忽略变量的占位符。

与 default 一起使用时，如果某个变量没有定义，那么使用 omit 占位符，Ansible 就会把这个对应的参数按照没有传这个参数的值来处理。

文件/tmp/foo 没有定义参数 mode，所以 default(omit)会在没有定义 mode 时忽略 mode 变量，Ansible 的 file 模块会按照没有传入 mode 这个参数来创建文件/tmp/foo。文件/tmp/baz 定义了 mode 为 0444，所以文件的权限为 0444。

```
- name: touch files with an optional mode
  file: dest={{item.path}} state=touch mode={{item.mode|default(omit)}}
  with_items:
    - path: /tmp/foo
    - path: /tmp/baz
      mode: "0444"
```

❸ mandatory：强制变量必须定义，否则抛错。

在 Ansible 默认的配置中，如果变量没有定义，那么直接使用未定义的变量{{some_undefined_variable}}会抛错。

如果在 Ansible 配置文件中使用了下面的配置，那么遇到未定义的变量时，Ansible 就不会抛错。

```
error_on_undefined_vars = False
```

而在此时如果想约束某一个变量必须定义，就可以使用 mandatory。

```
---
- hosts: localhost
  remote_user: root
  gather_facts: false

  vars:
    some_variable: "DefinedValue"

  tasks:
    #Please notice, the default setting of ansible is enforcing the variable to to defined
    #The following undefined variable is passed, just because in the ansible.cfg in this folder is overwrite the setting by:
    #error_on_undefined_vars = False
    - debug: msg="{{ some_variable }}"
    - debug: msg="{{ some_undefined_variable }}"
    - debug: msg="{{ some_undefined_variable | mandatory }}"
```

上面的例子在配置文件定义 error_on_undefined_vars = False 时的执行结果如下。

```
TASK [debug] ****************************************************
********
ok: [localhost] => {
    "msg": "DefinedValue"
}

TASK [debug] ****************************************************
********
ok: [localhost] => {
    "msg": "{{ some_undefined_variable }}"
}

TASK [debug] ****************************************************
********
fatal: [localhost]: FAILED! => {"failed": true, "msg": "Mandatory variable not defined."}
        to retry, use: --limit @/home/jshi/shared/learn_ansible/playbook_code/ansible-playbook-unclassified-topic/11_filter/site_filter_mandatory.retry
```

❹ bool：判断变量是否为布尔类型。

bool 和下节的 ternary 类型的过滤器都是 Playbook 中判断条件类型的过滤器。其中，bool 类型的过滤器用来判断变量是否为布尔类型。

举例说明：

```
---
- hosts: localhost
  remote_user: root
  gather_facts: false

  vars:
    some_string_value: "Test"
    some_book_value: True

  tasks:

    - debug: msg=test
```

```
      when: some_string_value | bool
    - debug: msg=test
      when: some_book_value | bool
```

执行的结果如下。

```
TASK [debug] ************************************************************
********
  skipping: [localhost]

TASK [debug] ************************************************************
********
ok: [localhost] => {
    "msg": "test"
}
```

❺ ternary：Playbook 的条件表达式。

ternary 类似于编程语言中的类型表达式，("A?B:C") 当条件为真时，返回前一个值；当条件为假时，返回另外一个值。

例如下面的代码。

```
  vars:
    name: "John"

  tasks:

    - name: "true or false take different value"
      debug: msg="{{ (name == "John") | ternary('Mr','Ms') }}"
```

执行的结果如下。

```
TASK [true or false take different value] *******************************
********
ok: [localhost] => {
    "msg": "Mr"
}
```

4.3.3　过滤器对文件路径的操作

Ansible 为了方便对文件及其路径进行操作，提供了一系列关于文件目录的操作，包含获取文件名、路径名，等等。因为 Linux 和 Windows 文件系统的路径名差异比较大，所以过滤器根据不同的系统，分别提供了不同类型的过滤器来处理。下面列举了一些常用的文件路径相关的过滤器。

Linux 文件路径的操作的过滤器如下。

- basename：获取路径中的文件名。
- dirname：获取文件的目录。
- expanduser：扩展~为实际的目录。
- realpath：获得链接文件所指文件的真实路径。
- relpath：获得相对某一根目录的相对路径。
- splitext：把文件名用点号（.）分割成多个部分。

上面的过滤器的用法都包含在下面的例子中。

```yaml
---
- hosts: localhost
  remote_user: root
  gather_facts: false

  vars:
    linux_path: "/etc/asdf/foo.txt"
    linux_user_path: "~/data/my_test/file.txt"
  tasks:
  #Path for linux system
  - name: "Get the filename in linux file path {{ linux_path }}"
    debug: msg="{{linux_path | basename}}"
  - name: "Get the dirname in linux file path {{ linux_path }}"
    debug: msg="{{linux_path | dirname}}"
  - name: "expanduser of {{linux_user_path}}"
    debug: msg="{{ linux_user_path | expanduser }}"
  - name: "realpath of {{ link_to_ansible_cfg }}"
    debug: msg="{{ 'link_to_ansible_cfg' | realpath }}"
  - name: "realpath of {{ link_to_ansible_cfg }}"
    debug: msg="{{ 'link_to_ansible_cfg' | relpath('/home') }}"
```

```
    - name: "Get the split name of file"
      debug: msg="{{ 'nginx.conf' | splitext}}"
```

执行的结果如下。

```
    TASK [Get the filename in linux file path /etc/asdf/foo.txt] *************
*******
    ok: [localhost] => {
        "msg": "foo.txt"
    }

    TASK [Get the dirname in linux file path /etc/asdf/foo.txt] *************
*******
    ok: [localhost] => {
        "msg": "/etc/asdf"
    }

    TASK [expanduser of ~/data/my_test/file.txt] ****************************
**********
    ok: [localhost] => {
        "msg": "/home/jshi/data/my_test/file.txt"
    }

    TASK [realpath of {{ link_to_ansible_cfg }}] ****************************
*********
    ok: [localhost] => {
        "msg":
"/home/jshi/shared/learn_ansible/playbook_code/ansible-playbook-unclassified
-topic/11_filter/ansible.cfg"
    }

    TASK [realpath of {{ link_to_ansible_cfg }}] *****************************
*******
    ok: [localhost] => {
        "msg": "jshi/shared/learn_ansible/playbook_code/ansible-playbook-
unclassified-topic/11_filter/link_to_ansible_cfg"
    }
```

```
TASK [Get the split name of file] *****************************
*********
ok: [localhost] => {
    "msg": "('nginx', '.conf')"
}
```

Windows 文件路径的操作的过滤器有：

◎ win_basename：获得 Windows 路径的文件名。
◎ win_dirname：获得 Windows 路径的文件目录。
◎ win_splitdrive：把 Windows 路径分割成多个部分。

在下面的例子中包含了对上面三个过滤器的用法。

```
vars:
  window_path: "C:\\window\\subfolder\\myfile.docx"
tasks:
  #Path for window system
  - name: "Get the filename in window file path {{ window_path }}"
    debug: msg="{{window_path | win_basename}}"
  - name: "Get the dirname in window file path {{ window_path }}"
    debug: msg="{{window_path | win_dirname}}"
  - name: "Get the split window file path {{ window_path }}"
    debug: msg="{{window_path | win_splitdrive}}"
  - name: "Get the split window file path {{ window_path }}"
    debug: msg="{{window_path | win_splitdrive|first}}"
  - name: "Get the split window file path {{ window_path }}"
    debug: msg="{{window_path | win_splitdrive|last}}"
```

执行的结果如下。

```
TASK [Get the filename in window file path C:\window\subfolder\myfile.docx]
****
ok: [localhost] => {
    "msg": "myfile.docx"
}

TASK [Get the dirname in window file path C:\window\subfolder\myfile.docx]
*****
ok: [localhost] => {
```

```
        "msg": "C:\\window\\subfolder"
    }

    TASK [Get the split window file path C:\window\subfolder\myfile.docx] **********
    ok: [localhost] => {
        "msg": "(u'C:', u'\\\\window\\\\subfolder\\\\myfile.docx')"
    }

    TASK [Get the split window file path C:\window\subfolder\myfile.docx] **********
    ok: [localhost] => {
        "msg": "C:"
    }

    TASK [Get the split window file path C:\window\subfolder\myfile.docx] **********
    ok: [localhost] => {
        "msg": "\\window\\subfolder\\myfile.docx"
    }
```

4.3.4 过滤器对字符串变量的操作

❶ quote：给字符串加引号。

```
    vars:
      my_test_string: "This is the test string"

    tasks:

      - name: "quote {{ my_test_string }}"
        debug: msg="echo {{ my_test_string | quote }}"
```

执行的结果如下。

```
    TASK [quote This is the test string] *************************************
    ok: [localhost] => {
```

```
    "msg": "echo 'This is the test string'"
}
```

❷ base64：得到字符串的 Base64 编码。

```
vars:
  my_comment: "This is the test comment"
  my_encoding_comment: "VGhpcyBpcyB0aGUgdGVzdCBjb21tZW50"

tasks:
  - name: "Get the base64 encoding string of {{ my_comment }}"
    debug: msg="{{ my_comment | b64encode }}"
  - name: "Get the base64 decoding string of {{ my_encoding_comment }}"
    debug: msg="{{ my_encoding_comment | b64decode }}"
  - name: "Get the uuid of {{ my_comment }}"
    debug: msg="{{ my_comment| to_uuid }}"
```

执行的结果如下。

```
TASK [Get the base64 encoding string of This is the test comment] **************
ok: [localhost] => {
    "msg": "VGhpcyBpcyB0aGUgdGVzdCBjb21tZW50"
}

TASK [Get the base64 decoding string of VGhpcyBpcyB0aGUgdGVzdCBjb21tZW50] ******
ok: [localhost] => {
    "msg": "This is the test comment"
}

TASK [Get the uuid of This is the test comment] ********************************
ok: [localhost] => {
    "msg": "1e854304-9b99-5811-9b41-2b835ceb9e11"
}
```

❸ hash：获取字符串的哈希值。

计算哈希值的算法有很多，如 sha1、md5、checksum 等。

```
vars:
```

```yaml
    my_password: "mytestpassword"

  tasks:
    - name: "Get sha1 hash of string {{ my_password }}"
      debug: msg="{{ my_password |hash('sha1') }}"
    - name: "Get md5 hash of string {{ my_password }}"
      debug: msg="{{ my_password |hash('md5') }}"
    - name: "Get checksum of string {{ my_password }}"
      debug: msg="{{ my_password |checksum }}"
    - name: "Get blowfish hash of string {{ my_password }}"
      debug: msg="{{ my_password |hash('blowfish') }}"
    - name: "Get sha512 password hash (random salt) of string {{ my_password }}"
      debug: msg="{{ my_password|password_hash('sha512') }}"
    - name: "Get  get a sha256 password hash with a specific salt of string {{ my_password }}"
      debug: msg="{{ my_password |password_hash('sha256', 'mysecretsalt') }}"
```

执行的结果如下。

```
    TASK [Get sha1 hash of string mytestpassword] ********************************
    ok: [localhost] => {
        "msg": "8f30637e207c02d9bd74da8e6ca13c6b8afbfe7b"
    }


    TASK [Get md5 hash of string mytestpassword] ********************************
    ok: [localhost] => {
        "msg": "7320d1035cb00f1e540bdbbb1d87ff69"
    }


    TASK [Get checksum of string mytestpassword] ***************************
    ok: [localhost] => {
        "msg": "8f30637e207c02d9bd74da8e6ca13c6b8afbfe7b"
    }


    TASK [Get blowfish hash of string mytestpassword] *********************
```

```
*******
    ok: [localhost] => {
        "msg": ""
    }

    TASK [Get sha512 password hash (random salt) of string mytestpassword]
*********
    ok: [localhost] => {
        "msg": "$6$eotF68dOmEx5EUCY$CvfElzkywbTLu7lRQya6NAoyb1CFkETz8LV6qy3MgpS9C1BuAxC5UHbx1FWt/mpsWySST1V9.szMiS4fJ4WDD0"
    }

    TASK [Get  get a sha256 password hash with a specific salt of string mytestpassword] ***
    ok: [localhost] => {
        "msg": "$5$mysecretsalt$lAIgb7Yl5zgQm3YT887bOwo8Fm.tgEujDe/k5UlPeND"
    }
```

❹ comment：把字符串变成代码注释的一部分。

comment 有很多风格和格式，在下面的例子中，展示了将字符串转化为不同风格和格式注释的使用方法，最后一个是用户自定义的注释风格。

```
    vars:
      my_comment: "This is the test comment"
    tasks:
      - name: "Simple comment"
        debug: msg="{{ my_comment | comment }}"
      - name: "C style comment"
        debug: msg="{{ my_comment | comment('c') }}"
      - name: "cblock style comment"
        debug: msg="{{ my_comment | comment('cblock') }}"
      - name: "erlang style comment"
        debug: msg="{{ my_comment | comment('erlang') }}"
      - name: "xml style comment"
        debug: msg="{{ my_comment | comment('xml') }}"
      - name: "customize style comment"
        debug: msg="{{ my_comment | comment('plain', prefix='#######\n#', postfix='#\n#######\n###\n#') }}"
```

执行的结果如下。

```
    TASK [Simple comment] *******************************************
*********
    ok: [localhost] => {
        "msg": "#\n# This is the test comment\n#"
    }

    TASK [C style comment] ******************************************
*******
    ok: [localhost] => {
        "msg": "//\n// This is the test comment\n//"
    }

    TASK [cblock style comment] *************************************
*******
    ok: [localhost] => {
        "msg": "/*\n *\n * This is the test comment\n *\n */"
    }

    TASK [erlang style comment] *************************************
*******
    ok: [localhost] => {
        "msg": "%\n% This is the test comment\n%"
    }

    TASK [xml style comment] ****************************************
*********
    ok: [localhost] => {
        "msg": "<!--\n -\n - This is the test comment\n -\n-->"
    }

    TASK [customize style comment] **********************************
*********
    ok: [localhost] => {
        "msg": "#######\n#\n# This is the test comment\n#\n#######\n###\n#"
    }
```

❺ regex：利用正则表达式对字符串进行替换。

```yaml
vars:
  my_comment: "This is the test comment"
tasks:
  - name: 'convert "ansible" to "able"'
    debug: msg="{{ 'ansible' | regex_replace('^a.*i(.*)$', 'a\\1') }}"
  - name: 'convert "foobar" to "bar"'
    debug: msg="{{ 'foobar' | regex_replace('^f.*o(.*)$', '\\1') }}"
  - name: 'convert "localhost:80" to "localhost, 80" using named groups'
    debug: msg="{{ 'localhost:80' | regex_replace('^(?P<host>.+):(?P<port>\\d+)$', '\\g<host>, \\g<port>') }}"
  - name: "# convert '^f.*o(.*)$'"
    debug: msg="{{ '^f.*o(.*)$' | regex_escape() }}"
```

执行的结果如下。

```
TASK [convert "ansible" to "able"] ****************************************
ok: [localhost] => {
    "msg": "able"
}

TASK [convert "foobar" to "bar"] ****************************************
ok: [localhost] => {
    "msg": "bar"
}

TASK [convert "localhost:80" to "localhost, 80" using named groups] ****************************************
ok: [localhost] => {
    "msg": "localhost, 80"
}

TASK [# convert '^f.*o(.*)$'] ****************************************
ok: [localhost] => {
    "msg": "\\^f\\.\\*o\\(\\.\\*\\)\\$"
}
```

❻ ip：判断字符串是否是合法的 IP 地址。

```
vars:
  my_valid_ip: "192.168.0.166"
  my_invalid_ip: "192.168.0.266"
  my_valid_ipv6: "fe80::dcfd:1e54:728d:a99e"
tasks:
  #Need to install python package netaddr to support ipaddr filter
  #pip install netaddr
  - name: "Check {{ my_valid_ip }} is an valid ip"
    debug: msg="{{ my_valid_ip | ipaddr }}"
  - name: "Check {{ my_invalid_ip }} is an valid ip"
    debug: msg="{{ my_invalid_ip | ipaddr }}"
  - name: "Check {{ my_valid_ip }} is an valid ipv4 addr"
    debug: msg="{{ my_valid_ip | ipv4 }}"
  - name: "Check {{ my_valid_ipv6 }} is an valid ipv6 addr"
    debug: msg="{{ my_valid_ipv6 | ipv6 }}"
  - name: "Check {{ '192.0.2.1/24' | ipaddr('address') }} is an valid ip"
    debug: msg="{{ '192.0.2.1/24' | ipaddr('address') }}"
```

执行的结果如下。

```
TASK [Check 192.168.0.166 is an valid ip] *****************************
*********
ok: [localhost] => {
    "msg": "192.168.0.166"
}

TASK [Check 192.168.0.266 is an valid ip] *****************************
********
ok: [localhost] => {
    "msg": false
}

TASK [Check 192.168.0.166 is an valid ipv4 addr] **********************
******
ok: [localhost] => {
    "msg": "192.168.0.166"
}
```

```
TASK [Check fe80::dcfd:1e54:728d:a99e is an valid ipv6 addr] ********
***********
ok: [localhost] => {
    "msg": "fe80::dcfd:1e54:728d:a99e"
}

TASK [Check 192.0.2.1 is an valid ip] ********************************
*********
ok: [localhost] => {
    "msg": "192.0.2.1"
}
```

❼ datetime：将字符串类型的时间转换成时间戳。

```
tasks:
  - name: "datetime filter"
    debug: msg="{{ (("2016-08-04 20:00:12"|to_datetime)
```

执行的结果如下。

```
TASK [datetime filter] ***********************************************
************
ok: [localhost] => {
    "msg": "72012"
}
```

4.3.5 过滤器对 JSON 的操作

❶ format：将变量的值按照 JSON/YAML 格式输出。

```
    - name: output the formated variable to /tmp/test_ansible_filter_
formated
      blockinfile:
        dest: /tmp/test_ansible_filter_formated
        block: |
          {{ some_variable | to_json }}
          -----------------------------
          {{ some_variable | to_yaml }}
```

```
                ------------------------------
                {{ some_variable | to_nice_json }}
                ------------------------------
                {{ some_variable | to_nice_yaml }}
```

在结果文件 **/tmp/test_ansible_filter_formated** 中的输出结果如下。

```
# BEGIN ANSIBLE MANAGED BLOCK
{"key2": "value2", "key1": "value1"}
------------------------------
{key1: value1, key2: value2}

------------------------------
{
    "key1": "value1",
    "key2": "value2"
}
------------------------------
key1: value1
key2: value2
```

❷ query 在一个 JSON 对象里，搜索符合条件的属性，返回符合条件的属性数组。

```
    vars:
      domain_definition:
        domain:
          cluster:
            - name: "cluster1"
            - name: "cluster2"
          server:
            - name: "server11"
              cluster: "cluster1"
              port: "8080"
            - name: "server12"
              cluster: "cluster1"
              port: "8090"
            - name: "server21"
              cluster: "cluster2"
              port: "9080"
            - name: "server22"
```

```yaml
                cluster: "cluster2"
                port: "9090"
        library:
          - name: "lib1"
            target: "cluster1"
          - name: "lib2"
            target: "cluster2"
```

在上述变量 domain_defination 中搜索出所有的 cluster 的名字。

```yaml
  - name: "Display all cluster names"
    debug: var=item
    with_items: "{{domain_definition|json_query('domain.cluster[*].name')}}"
```

搜索的结果如下。

```
TASK [Display all cluster names] ***************************************
************
ok: [localhost] => (item=cluster1) => {
    "item": "cluster1"
}
ok: [localhost] => (item=cluster2) => {
    "item": "cluster2"
}
```

❸ combine：合并两个 JSON 对象的值。

假设有三个变量。

```yaml
  vars:
    list1:
      key1: valueA1
      key2: valueA2
      key3: valueA3
      key4: valueA4
    list2:
      key1: valueB1
      key2: valueB2
      key3: valueB3
    list3:
      key1: valueC1
```

```
        key2: valueC2
    list4:
        key1: valueD1
```

对它们进行最简单的合并操作。

```
- name: "combine"
  debug: msg="{{ {'a':1, 'b':2}|combine({'b':3}) }}"
```

执行的结果为。

```
TASK [combine] ***************************************************
ok: [localhost] => {
    "msg": {
        "a": 1,
        "b": 3
    }
}
```

还可以进行深度递归。

```
- name: "combine with recursive"
  debug: msg="{{ {'a':{'foo':1, 'bar':2}, 'b':2}|combine({'a':{'bar':3, 'baz':4}}, recursive=True) }}"
```

4.3.6 过滤器对数据结构的操作

Ansible 中的过滤器支持以下几种类型的数据结构的操作。

❶ random：取随机数。

取随机数的操作比较常见，random 既支持从 List 中取随机的元素，也支持生成一个随机的数字。

取随机元素的操作。

```
- name: "List with random"
  debug: msg="{{ ['a','b','c']|random }}"
- name: "number with random"
  debug: msg="{{ 59 |random}} * * * * root /script/from/cron"
- name: "random with step"
```

```
      debug: msg="{{ 100 |random(step=10) }}"
    - name: "ramdom with start and step"
      debug: msg="{{ 100 |random(2, 10) }}"
    - name: "ramdom with start and step"
      debug: msg="{{ 100 |random(start=2, step=10) }}"
```

执行的结果如下。

```
TASK [List with random] *************************************************
ok: [localhost] => {
    "msg": "c"
}

TASK [number with random] ***********************************************
ok: [localhost] => {
    "msg": "41 * * * * root /script/from/cron"
}

TASK [random with step] *************************************************
ok: [localhost] => {
    "msg": "60"
}

TASK [ramdom with start and step] ***************************************
ok: [localhost] => {
    "msg": "72"
}

TASK [ramdom with start and step] ***************************************
ok: [localhost] => {
    "msg": "42"
}
```

此外，Ansible 的过滤器还支持对 List 和 Set 数据结构做如下操作。由于这些操作不是特别高频，因此这里仅仅列出其名字和描述，如果想了解其具体的语法，请参考链接：

https://github.com/ansible-book/ansible-playbook-unclassified-topic/tree/master/11_filter

❷ 对 List 的操作有：

- ◎ min：取最小值。
- ◎ max：取最大值。
- ◎ join：将 List 中的所有元素连接成一个新的字符串。
- ◎ shuffle：将 List 做顺序打乱成一个新的 List。
- ◎ map：实现对 List 的映射操作。

❸ 对 Set 的操作有以下过滤器：

- ◎ unique：去除重复的元素。
- ◎ union：取交集。
- ◎ differentce：取差集。
- ◎ symmetric_difference：取对称差。

4.3.7　过滤器的链式/连续使用

Ansible 的过滤器是支持链式使用的，即在一个{{}}中使用多个过滤器，就像下面这样。

```
- name: "Use multiple filter in chain"
  debug: msg="{{ [0,2] | map('extract', ['x','y','z']) | join('|') }}"
```

在上面的例子中，先用 map 操作得到['x', 'z']，再用 join 得到字符串'x|z'。

4.4　测试变量或表达式是否符合条件

测试和过滤器类似，是 Jinja2 提供的功能，同时也可以在 Ansible 中使用，使用的方法和过滤器类似。不同的是，测试的返回结果是 true 和 false，主要检查变量或表达式是否符合某一个条件。

Ansible 除了支持 Jinja2 自带的所有测试外，还提供了几个在 Ansible 中常用的测试功能。Jinja2 自带的测试，请参考下面的链接：

http://jinja.pocoo.org/docs/2.9/templates/#builtin-tests

4.4.1 测试字符串

match 和 search 是用于测试字符串是否符合某一个正则表达式的测试。其中 match 是完全匹配，search 只需部分匹配。

```
vars:
  url: "http://example.com/users/foo/resources/bar"

tasks:
  - debug: "msg='matched pattern 1'"
    when: url | match("http://example.com/users/.*/resources/.*")

  - debug: "msg='matched pattern 2'"
    when: url | search("/users/.*/resources/.*")

  - debug: "msg='matched pattern 3'"
    when: url | search("/users/")
```

上面例子的执行结果如下。

```
TASK [debug] *************************************************************
ok: [localhost] => {
    "msg": "matched pattern 1"
}

TASK [debug] *************************************************************
ok: [localhost] => {
    "msg": "matched pattern 2"
}

TASK [debug] *************************************************************
```

```
*******
ok: [localhost] => {
    "msg": "matched pattern 3"
}
```

4.4.2 比较版本

版本的比较是常用的功能，Ansible 提供了 version_compare 类型的测试来实现。

```
    vars:
      ansible_distribution_version: "11.05"

    tasks:
      - name: echo the remote version compare with 12.04
        debug: msg="{{ ansible_distribution_version | version_compare
('12.04', '>=') }}"
      - name: use version_compare with 3rd arguments
        debug: msg="{{ ansible_distribution_version | version_compare
('13.01', operator='lt', strict=True) }}"
```

执行的结果如下。

```
    TASK [echo the remote version compare with 12.04] ************************
******
    ok: [localhost] => {
        "msg": false
    }


    TASK [use version_compare with 3rd arguments] ****************************
*******
    ok: [localhost] => {
        "msg": true
    }
```

4.4.3 测试 List 的包含关系

issubset 可以分别测试 List 是否被包含或包含另外一个 List。

```
vars:
    a: [1,2,3,4,5]
    b: [2,3]
tasks:
  - debug: msg="A includes B"
    when: a|issuperset(b)

  - debug: msg="B is included in A"
    when: b|issubset(a)
```

执行的结果如下。

```
TASK [debug] ******************************************************************
ok: [localhost] => {
    "msg": "A includes B"
}

TASK [debug] ******************************************************************
ok: [localhost] => {
    "msg": "B is included in A"
}
```

4.4.4　测试文件路径

```
vars:
    mypath: /tmp
tasks:
  - debug: msg="path is a directory"
    when: mypath|is_dir

  - debug: msg="path is a file"
    when: mypath|is_file

  - debug: msg="path is a symlink"
    when: mypath|is_link
```

```
        - debug: msg="path already exists"
          when: mypath|exists

        - debug: msg="path is {{ (mypath|is_abs)|ternary('absolute', 'relative')}}"
```

执行的结果如下。

```
TASK [debug] ****************************************************************
ok: [localhost] => {
    "msg": "path is a directory"
}

TASK [debug] ****************************************************************
skipping: [localhost]

TASK [debug] ****************************************************************
skipping: [localhost]

TASK [debug] ****************************************************************
ok: [localhost] => {
    "msg": "path already exists"
}

TASK [debug] ****************************************************************
ok: [localhost] => {
    "msg": "path is absolute"
}
```

4.4.5　测试任务的执行结果

Ansible 提供了一系列的测试，可以用来判断任务的执行结果。

```
    tasks:
```

```yaml
    - shell: /usr/bin/foo
      register: result
      ignore_errors: True

    - debug: msg="it failed"
      when: result|failed

    # in most cases you'll want a handler, but if you want to do something right now, this is nice
    - debug: msg="it changed"
      when: result|changed

    - debug: msg="it succeeded in Ansible >= 2.1"
      when: result|succeeded

    - debug: msg="it succeeded"
      when: result|success

    - debug: msg="it was skipped"
      when: result|skipped
```

执行结果如下。

```
    TASK [command] *************************************************
*********
    fatal: [localhost]: FAILED! => {"changed": true, "cmd": "/usr/bin/foo", "delta": "0:00:00.002328", "end": "2017-02-12 06:57:41.013385", "failed": true, "rc": 127, "start": "2017-02-12 06:57:41.011057", "stderr": "/bin/sh: /usr/bin/foo: No such file or directory", "stdout": "", "stdout_lines": [], "warnings": []}
    ...ignoring

    TASK [debug] ***************************************************
*********
    ok: [localhost] => {
        "msg": "it failed"
    }
```

```
    TASK [debug] ****************************************************
********
    ok: [localhost] => {
        "msg": "it changed"
    }

    TASK [debug] ****************************************************
*******
    skipping: [localhost]

    TASK  [debug]  **************************************************
**********
    skipping: [localhost]

    TASK [debug] ****************************************************
********
    skipping: [localhost]
```

4.5 认识插件

Ansible 插件（plugin）并不像模块一样有统一出现的位置和相似的使用方法。它只是对 Ansible 功能的补充。Ansible 插件会因类型不同而使用不同的方法。如果想对上面提到过的 lookup 写更多的插件，使其功能更加丰富，那么应使用 lookup 插件，语法结构如下。

```
{{ lookup('new_lookup_plugin', "paramters"}}
```

而过滤器类型的插件，使用时的语法结构如下。

```
{{ variable | new_filter_plugin}}
```

在 Ansible 的配置文件中，每种类型的插件都有自己的配置变量，因此放置的目录并不相同。

4.5.1 插件类型

Ansible 除了 lookup 和过滤器外，还支持哪些类型的插件呢？请先看下面的插件配置变量列表，我们再来一一介绍。

```
# set plugin path directories here, separate with colons
#action_plugins     = /usr/share/ansible/plugins/action
#cache_plugins      = /usr/share/ansible/plugins/cache
#callback_plugins   = /usr/share/ansible/plugins/callback
#connection_plugins = /usr/share/ansible/plugins/connection
#lookup_plugins     = /usr/share/ansible/plugins/lookup
#inventory_plugins  = /usr/share/ansible/plugins/inventory
#vars_plugins       = /usr/share/ansible/plugins/vars
#filter_plugins     = /usr/share/ansible/plugins/filter
#test_plugins       = /usr/share/ansible/plugins/test
#strategy_plugins   = /usr/share/ansible/plugins/strategy
```

❶ Action 插件

和模块的使用方法类似，只不过执行目标不是远程主机，而是在 Ansible 的控制节点（管理节点）本机上。

❷ Cache 插件

为 Facts（主机变量）提供缓存，以避免多次执行 Playbook 时在搜集 Facts 上有过度的时间开销。

❸ Callback 插件

Ansible 执行 Playbook 后，提供额外的行为。例如，将执行结果发送到 E-mail 中，或者将执行结果写入 log 中，等等。

❹ connection 插件

Ansible 为管理节点和远程主机之间提供了更多的连接方法。默认的连接协议是基于 paramiko 的 SSH 协议。paramiko 对于大多数读者的基本需求已经够用，如果有高级的需求，则可以通过自定义的插件来提供 SNMP、Message bug 等传输协议。

❺ filters 插件

为前面提到的过滤器提供更多的功能。

❻ lookup 插件

为前面提到的 lookup 提供更多的功能。

❼ strategy 插件

为执行 Playbook 提供更多的执行策略，在后面会介绍 Ansible Strategy 的功能和用法。

❽ shell 插件

通过 shell 插件可以提供远程主机上更多类型的 shell（csh、ksh 和 tcsh）等的支持。

❾ test 插件

为前面提到的 Ansbile Jinja2 test 提供更多的功能。

❿ vars 插件

Ansible 可以将 inventory/playbook/命令行中的变量注入 Ansible 中。通过 vars 插件，可以实现更多方式的变量注入。

下面介绍几种常用的插件。

4.5.2 常用的插件介绍

1. lookup 插件

除了前面提到的 {{lookup()}} 语法外，像 "with_fileglob" 和 "with_items" 这样的使用方法，也是通过 lookup 插件来实现的。

2. filter 插件

和上面提到的过滤器的语法类似，是对过滤器功能的补充和增强。语法结构如下。

```
{{ variable | new_filter_plugin }}
```

3. strategy 插件

Ansbile strategy 控制 Play 在执行时的策略。默认的策略是 linear，即所有的远程主机都执行完某一个任务之后，再执行下一个任务。Ansible 官方的 strategy 插件提供了另外一种策略——free，语法如下，即允许每一个远程主机尽快地执行到 Play 的结尾。

```
- hosts: all
  strategy: free
  tasks:
  ...
```

4. callback 插件

callback 插件就是为 Playbook 执行后添加额外的行为，如果任务没有执行成功，则发送一封 E-mail，或是在执行后将执行结果放在 log 文件中。Ansible 官方提供了一些 callback 插件，具体请参考：

https://github.com/ansible/ansible/blob/devel/lib/ansible/plugins/callback

如果配置了 callball 插件，则 Ansible 会根据执行后的状态调用对应的 callback 插件。

下面通过一个完整的例子说明在 Ansbile 2.0 中如何使用已经写好的 callback 插件。

```
.
├── ansible.cfg
├── inventory.ini
├── plugins
│   └── callback
│       ├── __init__.py
│       ├── log_plays.py
│       ├── log_plays.pyc
│       ├── timer.py
│       └── timer.pyc
└── site_callback_plugin_example.yml
```

配置 Ansible callback 插件。

❶ 下载 callback plugins 文件

使用 callback 插件，首先从官网提供的 callback 插件中下载需要使用的插件对应的 Python 文件。本例中，将下载的文件保存在 plugins/callback 目录下。

这里以 timer 插件和 log_plays 插件为例，它们分别实现 Playbook 执行时间的计时，以及把每一个主机的执行结果记录在/var/log/ansible/hosts 对应主机名文件下。

❷ 配置 ansible.cfg

◎ 配置放置 callback 插件的 Python 文件的位置。
◎ 在所有的 callback 插件中，你想使用哪些 callback 插件。

配置文件的实例如下。

```
callback_plugins = plugins/callback
callback_whitelist = timer, log_plays
```

Playbook 文件 site_callback_plugin_example.yml 中不需要添加任何额外的设置和语句，Ansible 会自动调用 callback plugin timer 和 log_plays。

```
---
- hosts: all
  remote_user: root
  gather_facts: false

  tasks:
  - name: Test callback plugin
    shell: ls /tmp
```

执行的结果如下。

```
[jshi@jjshi 04_plugin]$ ansible-playbook site_callback_plugin_example.yml

PLAY [all] *********************************************************************

TASK [Test callback plugin] ****************************************************
changed: [localhost]
changed: [jshi-test-02.rhev]

PLAY RECAP *********************************************************************
jshi-test-02.rhev          : ok=1    changed=1    unreachable=0    failed=0
localhost                  : ok=1    changed=1    unreachable=0    failed=0
```

```
Playbook run took 0 days, 0 hours, 0 minutes, 1 seconds
Playbook run took 0 days, 0 hours, 0 minutes, 1 seconds
```

在执行结果的最后,输出了 timer 提供的 Playbook 的执行时间。

再去 /var/log/ansible/hosts 目录下验证记录的 log,出现了 host 名字对应的文件,文件里面记录的内容由多条 JSON 组成,如下所示。

```
Feb 11 2017 20:43:54 - OK - {"module_name": "command", "module_args": {"warn":
true, "executable": null, "_uses_shell": true, "_raw_params": "ls /tmp",
"removes": null, "creates": null, "chdir": null}} => {"_ansible_parsed": true,
"cmd": "ls /tmp", "end": "2017-02-11 20:43:54.773784", "_ansible_no_log": false,
"stdout": "ansible_b1CnkY\nplugtmp\nsystemd-private-05e1ed9540ad433e9645aa0e
8857cfad-colord.service-yR2lco\nsystemd-private-05e1ed9540ad433e9645aa0e8857
cfad-rtkit-daemon.service-qYWunA\ntest\ntracker-extract-files.1000", "changed":
true, "start": "2017-02-11 20:43:54.770475", "delta": "0:00:00.003309", "stderr":
"", "rc": 0, "stdout_lines": ["ansible_b1CnkY", "plugtmp", "systemd-private-
05e1ed9540ad433e9645aa0e8857cfad-colord.service-yR2lco",  "systemd-private-
05e1ed9540ad433e9645aa0e8857cfad-rtkit-daemon.service-qYWunA", "test", "tracker-
extract-files.1000"], "warnings": []}
```

第 5 章

role 和 Ansible Galaxy

本章重点

5.1　role 和 Ansible Galaxy 的简要介绍
5.2　role 的放置位置
5.3　在 Playbook 中如何调用 role
5.4　如何写 role
5.5　role 的依赖
5.6　Ansible Galaxy 网站介绍
5.6　演示 role 的创建和分享

5.1 role 和 Ansible Galaxy 的简要介绍

5.1.1 role

role 是高级版本的 include 语句，include 用来分享单个 Playbook 文件，而 role 可以分享一个文件夹，文件夹里面包含 task（任务）、handlers、file（文件）、template（模板）和 variable（变量）。

文件夹里面包含了实现一个完整的功能所有文件，例如，安装和配置 Web 服务器 Nginx、安装和配置数据库 MySQL 等，Ansible 中的 role 与编程语言中包的概念类似。

5.1.2 Ansible Galaxy

Ansible Galaxy 是 Ansible 提供的分享 role 的网站。

5.2 role 的放置位置

role 文件夹都可以放在哪里呢？除了当前子目录 roles 下，可不可以放在其他文件夹下，以便让更多 Playbook 项目分享同一组 role 呢？当然可以。下面就来总结 role 的放置位置。

5.2.1 当前目录的 roles 文件夹下

无论 Ansible 中对 roles path 是如何设置的，放在当前子目录 roles 文件夹下的 role 都会被找到。

5.2.2　环境变量 ANSIBLE_ROLES_PATH 定义的文件夹

如果定义了环境变量 ANSIBLE_ROLES_PATH，那么 Ansible 也会搜索该文件夹下放置的 role。

5.2.3　Ansible 配置文件中 roles_path 定义的文件夹

在前面的章节中，我们讲到 Ansbile 的配置文件中可以自定义很多变量，使得 Ansible 的使用更加灵活。其中，变量 roles_path 就允许用户自定义放置 role 的文件夹。

我们先回顾一下 Ansible 的配置文件都有哪些。

```
* file that ANSIBLE_CONFIG (an environment variable) pointed to
* ansible.cfg (in the current directory)
* .ansible.cfg (in the home directory)
* /etc/ansible/ansible.cfg
```

定义配置变量 roles_path 的格式如下，如果有多个目录，则使用冒号（:）分隔。

```
roles_path = /opt/mysite/roles:/opt/othersite/roles
```

注意，让配置变量 roles_path 生效是有条件的！如果没有定义环境变量 ANSIBLE_ROLES_PATH，那么还可以通过配置变量 roles_path 定义的文件夹。但是如果已经定义了环境变量 ANSIBLE_ROLES_PATH，那么配置变量 roles_path 定义的文件夹就会失效。如果已经定义了环境变量，并且想测试配置变量 **roles_path**，则可以通过下面的命令取消环境变量的设置。

```
unset ANSIBLE_ROLES_PATH
```

5.2.4　默认文件夹/etc/ansible/roles

如果既没有定义环境变量 ANSIBLE_ROLES_PATH，也没有在配置文件中定义变量 roles_path，那么 Ansible 还提供了一个默认的文件夹——/etc/ansible/roles，其中放置的 role 可以在所有 Playbook 中使用。

5.3 在 Playbook 中如何调用 role

假设有一个最简单的 role 叫作 simple，里面只有一个任务（后面会介绍完整的 role 的定义），那么在 Playbook 中如何使用 role 呢？

5.3.1 调用最简单的 role

调研 role simple 的 Playbook 文件是 site.yml，通过关键字 roles 来调用 role。Ansible 会自动到当前 Playbook 所在目录的子目录 roles 下查找对应名字的 role。

```yaml
---
#文件 site.yml
- hosts: all
  remote_user: root
  roles:
    - simple
  tasks:
    - name: Tasks in site.yml
      debug: msg="This is a task in site.yml"
```

role simple 中的内容和文件目录如下。

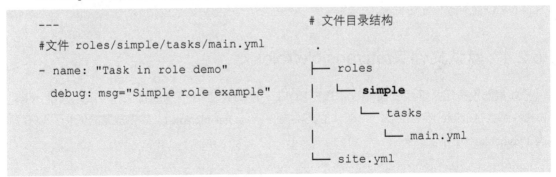

文件 site.yml 的执行结果如下。

Ansible 会先执行 role，再执行本文件中包含的任务。

```
PLAY [all] ****************************************************************
********
TASK [setup] **************************************************************
********
ok: [localhost]

TASK [simple : Task in role simple] ***************************************
**********
ok: [localhost] => {
    "msg": "Simple role example"
}

TASK [Tasks in site.yml] **************************************************
******
ok: [localhost] => {
    "msg": "This is a task in site.yml"
}

PLAY RECAP ****************************************************************
********
localhost                  : ok=3    changed=0    unreachable=0    failed=0
```

5.3.2 通过 pre_tasks 和 post_tasks 调整 role 和任务的顺序

如果想让一些任务在 role 之前和之后执行，也可以通过 Ansible 提供的另外两个关键字 pre_tasks 和 post_tasks 来实现。

在 Playbook 文件 site.yml 里面的使用方法如下。

```
---
- hosts: all
  remote_user: root

  pre_tasks:
    - name: pre task
```

```
      shell: echo 'hello' in pre_tasks
  roles:
    - simple
  tasks:
    - name: Tasks in site*.yml
      debug: msg="This is a task in site*.yml"
  post_tasks:
    - name: post task
      shell: echo 'goodbye' in post_tasks
```

执行的结果如下。

```
$ ansible-playbook site_with_pre_and_post.yml

PLAY [all] *********************************************************************

TASK [setup] *******************************************************************
ok: [localhost]

TASK [pre task] ****************************************************************
changed: [localhost]

TASK [simple : Task in role demo] **********************************************
ok: [localhost] => {
    "msg": "This is very simplest role without any variable"
}

TASK [Tasks in site*.yml] ******************************************************
ok: [localhost] => {
    "msg": "This is a task in site*.yml"
}

TASK [post task] ***************************************************************
```

```
*********
    changed: [localhost]

PLAY RECAP **********************************************************
********
localhost    : ok=5    changed=2    unreachable=0    failed=0
```

5.3.3 调用带有参数的 role

调用带有参数的 role 有两种方法：一是把 role 写成 JSON Object 的格式，直接传入参数；二是通过 vars 关键字使用 YAML 字典格式传入关键字。

```
---
#pass variable to role
- hosts: all
  remote_user: root

  roles:
    - { role: my_role_with_vars, ex_param: "Hello from ex_param for the 1st time"}
    - role: my_role_with_vars
      vars:
        ex_param: "Hello from ex_param for the 2nd time"
        in_param1: "This is the value from external"
```

5.3.4 与 when 一起使用 role

role 当然也可以和 when 一起使用，即当满足一定条件时再执行 role。与 role 的传参语法类似，调用方式有两种。

```
---
#Use role with condition
- hosts: all
  remote_user: root

  roles:
    - { role: simple, when: "ansible_os_family == 'Dibian'" }
```

```
    - role: simple
      when: "ansible_os_family == 'Dibian'"
```

5.4 如何写 role

5.4.1 role 的完整定义

什么是 role?

- role 就是更高级版本的 include 语句，include 语句只能重用一个包含任务的 Playbook 文件，而 role 可以分享和重用一个包含多种文件的文件夹，其中，文件可以是任务、handler、静态文件、模板文件以及定义变量的文件等。
- role 的实现机制并不复杂，只是自动的加载（include）一些文件，并提供一些自动搜索的文件路径。

role 是通过遵循特定的文件夹的结构来定义的，我们定义了一个名字为 x 的 role。

- 如果文件 role/x/taks/main.yml 存在，则文件中列出的任务都会被加入 Play 中。
- 如果文件 role/x/handlers/main.yml 存在，则文件中列出的 handlers 都会被加入 Play 中。
- 如果文件 role/x/vars/main.yml 存在，则文件中列出的变量都会被加入 Play 中。
- 如果文件 role/x/defaults/main.yml 存在，则文件中列出的变量都会被加入 Play 中。
- 如果文件 role/x/meta/main.yml 存在，则文件中列出的所有依赖的 role 都会被加入 Play 中。
- 此外，下面的文件不需要绝对或者是相对路径，和放在同一个目录下的文件一样，直接使用即可。
    - copy 或者 script 使用 roles/x/files/ 下的文件。
    - template 使用 roles/x/templates 下的文件。
    - include 使用 roles/x/tasks 下的文件。

一个包含完整文件结构的 role 如下面的图所示，这里只是列出了所有可能被自动加载的文件，这里面任何一个文件都不是必需的。如果文件不存在，则会跳过对该文件的加载。我们在"Ansible Role 的调用方法"中曾用到一个只包含 tasks/main.yml 文件的 role simple。

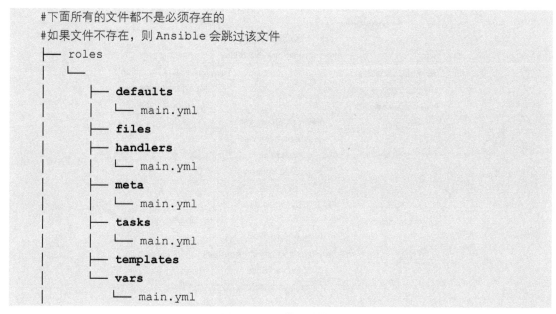

图 5.1 role 的文件结构

上面的目录结构除了手动创建外,Ansible 还提供了一个命令来创建 role 所需的完整的目录结构。

```
ansible-galaxy init role_name
```

5.4.2 默认变量和普通变量的区别

在前面的定义中,你也许会感到好奇,为什么文件夹 defaults 和 vars 下面的变量都会加到 Play 中,它们有什么区别吗?

defaults/main.yml 中的变量是默认变量。优先级在所有的变量中是最低的,用于放置一些需要被覆盖的变量。

vars/main.yml 中的变量是 role 变量,优先级比较高,放置一些不想被覆盖的变量,所以变量在命名的时候一般都加入 role 的名字作为前缀,防止不小心被 Playbook 中定义的变量覆盖。

关于 Ansible 中变量的优先级,可以简单地参考图 5.2。

Ansible	Tower
role defaults	
dynamic inventory variables	
inventory variables	Tower inventory variables
inventory group_vars	Tower group variables
inventory host_vars	Tower host variables
playbook group_vars	
playbook host_vars	
host facts	
registered variables	
set_facts	
play variables	
play vars_prompt	(not supported in Tower)
play vars_files	
role variables and include variables	
block variables	
task variables	
extra variables	Job Template extra variables Job Template Survey (defaults) Job Launch extra_vars

图 5.2 Ansible 中变量的优先级

5.4.3　tasks/main.yml 如何使用变量、静态文件和模板

任务是 Playbook 及 role 的核心逻辑。所以要想知道 role 做了什么，首先要看 role 的任务文件 tasks/main.yml，该文件也可以看作是一个 role 的入口文件。学会如何在该文件中使用变量、静态文件和模板等资源是学会写 role 的关键。

role 中的资源可以分为两类：一类是放在 x/*/main.yml 中会自动被加载的资料；还有一类是放在文件 x/*/other_but_main.yml 中，需要显式调用的资源。

下面介绍如何使用这两类资源。

1. 使用 x/*/main.yml 中的变量和 handler

❶ 使用 x/*/main.yml 中的变量和 handler

◎　像使用在同一个 Playbook 中的资源一样使用 x/*/main.yml 中的变量和 handlers

❷ 使用 x/{files,templates}/下的文件。

◎ 像使用同一个目录下的文件一样来使用放在这里的变量。当然，要求模块和使用的文件类型要匹配。

— copy、script 对应 files 目录下的文件。

— template 对应 templates 目录下的文件。

比如，role x 的目录结构如图 5.3 所示。

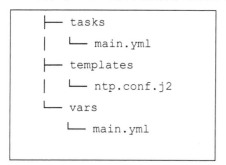
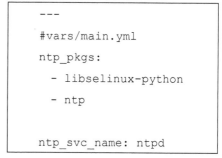

图 5.3 role x 的目录结构

在 tasks/main.yml 中使用变量和模板文件的方法如下。

```
---
#x/*/main.yml
- name: Install the required packages in Redhat
  yum: name={{ item }} state=installed
  with_items: '{{ ntp_pkgs }}'
  when: ansible_os_family == 'RedHat'

- name: Copy the ntp.conf template file
  template: src=ntp.conf.j2 dest=/etc/ntp.conf
```

2. 使用 x/*/other_but_main.yml 中的资源

如果 role x 下面的内容比较复杂，需要对任务或者 vars 进一步分类，可以使用除 main.yml 以外的文件外。应如何使用其他文件中的任务或者 vars 呢？Ansible 提供了两个关键字：include 和 include_vars，来分别引入 role 中非 main.yml 其他文件包含的任务和 vars。

例如，在下面的 role 中，把 RedHat Linux 和 Debian Linux 分别需要的变量放到不同的变量文件中，需要的任务分别放在不同的任务文件中。

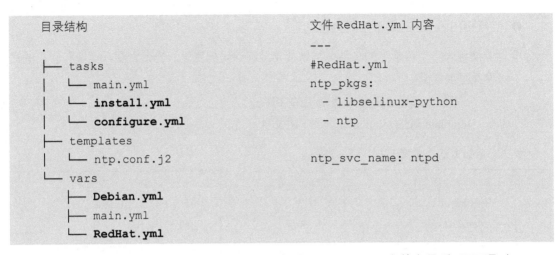

那么，在 x/*/main.yml 中，通过 include_vars 来引入 RedHat.yml 中的变量后，即可通过 include 来加载文件 install.yml 和 configre.yml 中的任务。

```
---
#myrole/*/main.yml
- name: Add the OS specific varibles
  include_vars: RedHat.yml
- name: Install the required packages in Redhat
  yum: name={{ item }} state=installed
  with_items: '{{ ntp_pkgs }}'
  when: ansible_os_family == 'RedHat'

- include: install.yml
- include: configure.yml
```

5.5 role 的依赖

安装一个 Nginx 需要配置 yum/apt 仓库，如果不想在配置 Nginx 的 Playbook 重新实现配置 yum/apt 仓库的功能，那么就可以通过 role 的依赖来解决。role 依赖关系的定义文件是 x/meta/main.yml。如果在 role x 定义依赖 role y，那么在 Playbook 中调用 role x 之前会先调用 role y。当多个 role 依赖同一个 role 时，Ansible 会自动进行过滤，避免重复调用相同参数的 role。

在下面的例子中，role db 和 Web 都依赖 role common。如果在 Playbook 中调用 db 和 Web，那么 Ansible 会保证在 role db 和 Web 运行前，先运行 role common，并且只运行一次。

```
├── roles
│   ├── common
│   │   └── tasks
│   │       └── main.yml
│   ├── db
│   │   ├── meta
│   │   │   └── main.yml
│   │   └── tasks
│   │       └── main.yml
│   └── Web
```

```
---
#在{web,db}/meta/main.yml 中
dependencies:
- { role: common }
```

```
#在 site.yml 文件中
…
roles:
# Role 的 depency 相当于隐形地加入了 common
# - common
  - db
  - web
```

执行结果如下。

```
$ ansible-playbook site.yml

PLAY [apply common configuration to all nodes] ********************************

TASK [setup] ******************************************************************
ok: [localhost]

TASK [common : Task in role common] *******************************************
ok: [localhost] => {
    "msg": "No variable is required in this role"
```

```
    }

    TASK [db : Task in role db] *********************************************
    ******
    ok: [localhost] => {
        "msg": "This is the task of db"
    }

    TASK [web : Task in role web] *******************************************
    ********
    ok: [localhost] => {
        "msg": "This is the task in role web"
    }

    PLAY RECAP **************************************************************
    ********
    localhost                  : ok=4    changed=0    unreachable=0    failed=0
```

当然，依赖关系中和 role 的调用一样，也是可以加入参数的，下面是加入参数的 meta/main.yml 文件的例子。在 Ansible 的去重机制中，只有对一个相同的 role 进行参数相同的调用时，才算是重复的。

```
---
dependencies:
- { role: common, name: "NameInDb" }
```

5.6 Ansible Galaxy 网站介绍

5.6.1 从 Ansible Galaxy 网站上下载 role

首先找到哪个 role 是你想要的。

访问 https://galaxy.ansible.com/，通过 BROWSE ROLES 找到你需要的 role。

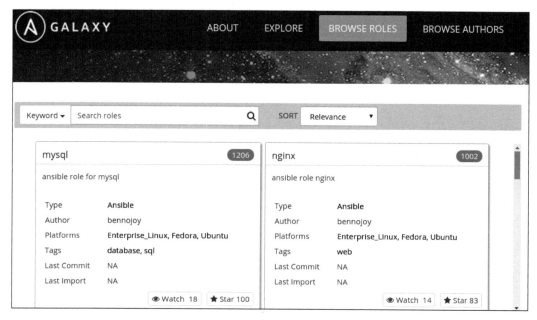

图 5.4　Galaxy 主页

每个 role 的详情页面都会列出下载该 role 的命令。例如，在如图 5.5 所示的截图中，role bennojoy.nginx 的安装命令已在该 role 的详情页面的 Installation 一行中列出。

图 5.5　role bennojoy.nginx 的详情页面

1. 一次下载一个 role

Ansible 提供了 ansible-galaxy install 命令来下载 role 到本地。参数 -p 指定 role 下载的目录。如果没有指定 -p，那么 role 会被自动下载到环境变量 ANSIBLE_ROLES_PATH 定义的目录下，或者默认目录 /etc/ansible/roles 下。

```
ansible-galaxy install -p ./roles username.rolename
```

2. 一次下载多个 role

将所有依赖的 role 放在一个 role.txt 里面，然后 -r 参数指定包含批量下载 role 的列表文件 role.txt。

批量下载的命令。

```
$ ansible-galaxy install -r roles.txt
```

roles.txt 样例。

```
user1.role1,v1.0.0
user2.role2,v0.5
user2.role3
```

如果不满足于只从 Ansible Galaxy 中下载，那么还可以在 YAML 文件中，使用 src、name 等参数来实现从其他网站中下载 role。

批量下载的命令。

```
$ ansible-galaxy install -r requirements.yml
```

requirements.yml 文件的样例。

```
# from galaxy
- src: yatesr.timezone

# from GitHub
- src: https://github.com/bennojoy/nginx

# from GitHub, overriding the name and specifying a specific tag
- src: https://github.com/bennojoy/nginx
  version: master
  name: nginx_role
```

```
# from a webserver, where the role is packaged in a tar.gz
- src: https://some.webserver.example.com/files/master.tar.gz
  name: http-role
```

5.6.2 分享你的 role

分享一个 role 可分为非常简单的三步，具体如下。

- 在你的 Github 账户下为你需要分享的 role 单独创建一个 Gitpub 的项目。
- 用 Github 账户登录 Ansbile Galaxy。
- 到 MY ROLES 菜单下，单击 REFRESH，GALAXY 就会搜索你 Github 账户下所有的项目，并且将符合 role 目录结构的所有项目列出来。在列表中，有开启/关闭分享 role 菜单的功能，根据你的需要，开启 role 的分享就可以了。

5.7 演示 role 的创建和分享

让我们创建一个 role 并通过 Ansible Galaxy 来分享它吧！

下面用一个完整的单个 Playbook 来实现 RedHat7/CentOS7 上 Nginx 的安装和配置。我们把它转化成一个可以被分享和重用的 role，然后通过 Ansible Galaxy 来分享我们写好的 role。

```
---
- hosts: all
  remote_user: root
  vars:
    nginx_yum_repo_enabled: true
    nginx_package_name: "nginx"
    nginx_vhost_path: /etc/nginx/conf.d
    nginx_vhosts_filename: "vhosts.conf"
    nginx_vhosts:
      - listen: 3101
        root: "/var/www/nginx3101"
        index: "index.html index.htm"
    firewall_disable_firewalld: false
```

```yaml
tasks:
  - name: Enable nginx repo.
    template:
      src: templates/nginx.repo.j2
      dest: /etc/yum.repos.d/nginx.repo
      owner: root
      group: root
      mode: 0644
    when: nginx_yum_repo_enabled
  - name: Ensure nginx is installed.
    yum:
      name: "{{ nginx_package_name }}"
      state: installed
  - name: Close SELinux
    command: setenforce 0
  - name: Ensure vhost.root exists.
    file:
      path: "{{ item.root }}"
      state: directory
    with_items: "{{ nginx_vhosts }}"
  - name: Copy the index.html for vhost
    copy:
      src: files/index.html
      dest: "{{ item.root }}/index.html"
      mode: 0644
    with_items: "{{ nginx_vhosts }}"
  - name: Add managed vhost config file (if any vhosts are configured).
    template:
      src: templates/vhosts.j2
      dest: "{{ nginx_vhost_path }}/{{ nginx_vhosts_filename }}"
      mode: 0644
    when: nginx_vhosts|length > 0
    notify: reload nginx
  - name: Ensure nginx is started and enabled to start at boot.
    service: name=nginx state=started enabled=yes
  - firewalld:
      port: "80/tcp"
```

```yaml
      permanent: true
      state: enabled
    when: firewall_disable_firewalld == false
    notify: restart firewalld
  - firewalld:
      port: "{{item.listen}}/tcp"
      permanent: true
      state: enabled
    with_items: "{{ nginx_vhosts }}"
    when: firewall_disable_firewalld == false
    notify: restart firewalld
  - name: Disable the firewalld service
    service:
      name: firewalld
      state: stopped
      enabled: no
    when: firewall_disable_firewalld
handlers:
  - name: restart nginx
    service: name=nginx state=restarted
  - name: reload nginx
    service: name=nginx state=reloaded
  - name: restart firewalld
    service: name=firewalld state=restarted
```

5.7.1 改造单个的 Playbook 为 role

role 的名字定为 nginx_on_redhat。

第一步，把 handlers 放到 nginx_on_redhat/handlers/main.yml 中，files 下面的文件放到 nginx_on_redhat/files/文件夹下面，templates 下面的文件放到 nginx_on_redhat/templates/文件夹下面。

第二步，根据变量是否想被覆盖，分别把变量放到 defaults/main.yml 和 vars/main.yml 中。

是否需要配置 yum 仓库、vhosts 的端口号以及防火墙信息是需要用户根据自己的实际情况判定的，所以要放在 defaults/main.yml 下，以方便用户覆盖这些变量的值。

```yaml
#defaults/main.yml
---
nginx_yum_repo_enabled: true
nginx_vhosts:
  - listen: 3103
    root: "/var/www/nginx3103"
    index: "index.html index.htm"
firewall_disable_firewalld: false
```

Nginx 配置文件的位置、配置文件的名字以及 Nginx 的安装包的名字是相对固定的，可以直接放在 vars/main.yml 中。不过这些值在 Debian 系统和 RedHat 系统中差别很大，为了方便这个 role 以后能扩展到 Debian 系统，我们把这些变量放在 vars/RedHat.yml 中。

```yaml
---
nginx_package_name: "nginx"
nginx_vhost_path: /etc/nginx/conf.d
nginx_vhosts_filename: "vhosts.conf"
```

第三步，把所有的任务放在 tasks/main.yml 下面，并注意修改 copy 模块和 templates 模块中的文件路径。

下面为修改文件路径前的 Copy 模块和 template 模块。

```yaml
- name: Copy the index.html for vhost
  copy:
    src: files/index.html
    dest: "{{ item.root }}/index.html"
    mode: 0644
  with_items: "{{ nginx_vhosts }}"
- name: Add managed vhost config file (if any vhosts are configured).
  template:
    src: templates/vhosts.j2
    dest: "{{ nginx_vhost_path }}/{{ nginx_vhosts_filename }}"
    mode: 0644
  when: nginx_vhosts|length > 0
  notify: reload nginx
```

由上面的带文件路径的引用，改为下面的不需要文件路径的引用。

```yaml
- name: Copy the index.html for vhost
  copy:
```

```
    src: index.html
    dest: "{{ item.root }}/index.html"
    mode: 0644
  with_items: "{{ nginx_vhosts }}"
- name: Add managed vhost config file (if any vhosts are configured).
  template:
    src: vhosts.j2
    dest: "{{ nginx_vhost_path }}/{{ nginx_vhosts_filename }}"
    mode: 0644
  when: nginx_vhosts|length > 0
  notify: restart nginx
```

第四步，在 tasks/main.yml 中开始加载 vars/RedHat.yml 中的变量。

```
---

- name: Add the OS specific varibles
  include_vars: "RedHat.yml"
```

第五步，为了进一步提高 role 的可读性，我们将 tasks/main.yml 冗长的任务列表，根据其功能，分别放在 tasks/{install.yml,selinux.yml,vhosts.yml,serivce.yml,firewalld.yml} 中，并在 tasks/main.yml 中加载这些文件。

```
---
#完整的tasks/main.yml
- name: Add the OS specific varibles
  include_vars: "RedHat.yml"
- include: install.yml
- include: selinux.yml
- include: vhosts.yml
- include: service.yml
- include: firewalld.yml
```

让我们来测试一下这个 role 吧！在 Playbook 的根目录中创建 Playbook 文件 site.yml 来调用 roles 的 nginx_on_redhat。

```
---

- hosts: all
  remote_user: root
```

```
vars:
  nginx_vhosts:
    - listen: 3104
      root: "/var/www/nginx3104"
      index: "index.html index.htm"
roles:
  - nginx_on_redhat
```

执行命令。

ansible-playbook site.yml

测试 Playbook 的执行结果，访问网站 http://<remoteserver>:3001/。

5.7.2 在 Ansible Galaxy 中分享 role

❶ 在 Github 中创建项目 nginx_on_redhat

将本地的 role 文件夹 nginx_on_redhat 中的代码，上传代码到新的 Github 项目中。

```
git init
git remote add origin git@github.com:shijingjing1221/nginx_on_redhat.git
git commit -am ""
git push -u origin master
```

❷ 用 Github 账户登录 Ansible Galaxy 网站

单击右上角的 MY ROLES 菜单，单击"刷新"按钮并安装，Galaxy 会自动识别出新创建的 role 项目。单击该 role 名字上的链接，会跳转到 role 的详情页面，就可以看见如何下载该 role 的命令啦。下面是新创建 role 的下载命令。

```
$ ansible-galaxy install shijingjing1221.nginx_on_redhat
```

第 6 章

Ansible Tower

本章重点

6.1 为什么要用 Ansible Tower
6.2 如何使用 Ansible Tower
6.3 与第三方平台的整合

6.1 为什么要用 Ansible Tower

6.1.1 Ansible 和 Tower 的用户视角架构图

在企业中，被管理的远程主机有成百上千台，每位管理员都需要配置所有主机的 SSH 连接，无疑工作量是巨大的。如果每位管理员的电脑上都存储了 SSH 密钥，那么一旦管理员的工作出现变动，无疑会出现很多的安全隐患，尤其是线上产品使用的主机，如图 6.1 所示。

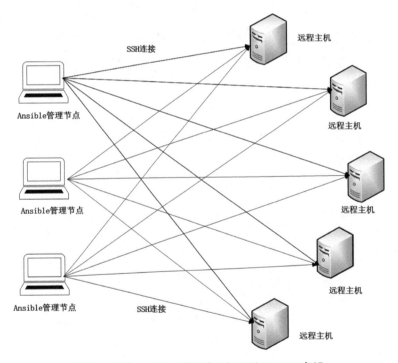

图 6.1 每个 Ansible 管理节点都存储了 SSH 密钥

Ansible Tower 解决了企业级用户面临的上述难题，它是中心化的 Ansible 管理节点，远程主机管理员通过登录 Tower 来运行 Playbook，无须每位管理员都在自己的机器上配置 SSH 连接，提高了工作效率。另外，在 Tower 中有 SSH 使用权限的访问控制，即便有权限使用 SSH 秘钥的管理员，也不能直接看到 SSH 密钥的明文，从而确保了安全。Tower 的架构图如图 6.2 所示。

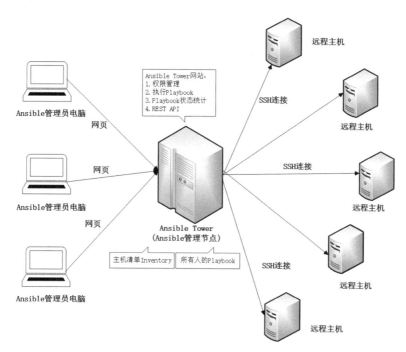

图 6.2　Tower 的架构图

6.1.2　Ansible Tower 的主要功能

Ansible Tower 作为一个中心化的 Ansible 管理节点网站，它自然会包含两部分功能：

❶ 用户和权限的管理。

❷ Ansible 命令行工具能做的事，通过 Tower 网站图形化界面都可以做。

此外，为了提高用户的体验，Tower 还提供了另外两个功能：

❶ 记录 Playbook 的所有执行结果，提供用户视角和主机视角的统计信息。

❷ 提供了 API 接口，让第三方平台可以根据 API 调用 Tower 的功能。

当然，Tower 最重要的功能就是如何通过图形化的界面来实现 Ansbile 命令行工具能实现的功能，下面我们重点介绍。

首先回顾一下 Ansible 命令行工具的使用工程，然后再对比 Tower，就会清楚 Tower 是如何使用的了。

在 Ansible 命令行工具中：

◎ 需要配置基于密钥的 SSH 连接。
◎ 把新的远程主机加到 inventory 中。
◎ 准备 Playbook 及其需要的所有文件。
◎ 运行 Playbook。

Ansible Tower 的主要功能和 Ansible 命令行的使用相对应。

首先来看看 Tower 网站的主界面。

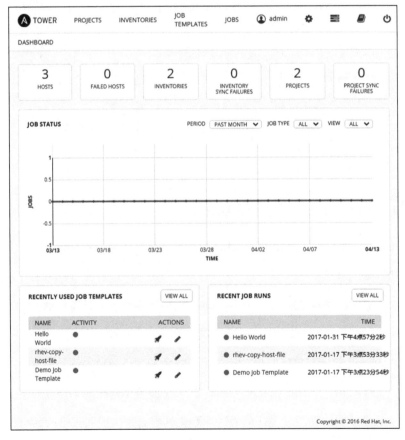

图 6.3　Tower 网站的主界面

左上角有四个菜单：PROJECTS、INVENTORYS、JOB TEMPLATES 和 JOBS，这就是 Tower 的主要功能。它们的含义如下。

❶ PROJECT：放置 Playbook 的文件夹，可以放在 Tower 网站的本地，更常见的是放在 git/svn 的服务器上。

❷ INVENTORIES：和命令行工具中的含义一样，配置需要管理的远程主机的列表和分组。

❸ JOB TEMPLATES：选择执行一次 Playbook 所需要的所有的信息，包括如下内容。

- ◎ 选择 credential。
  - — SSH 连接需要的密码或者密钥。
  - — 因为所有用户都需要配置 credential，所以 Tower 把创建 credential 的功能放在 setting 里面了。
- ◎ 选择 inventory。
- ◎ 选择 project。
  - — 选择 project 中执行哪个 YAML 文件。

❹ JOBS：记录了每一次 Playbook 执行的结果。

- ◎ 查看实时更新的执行结果。
- ◎ 查看所有执行的历史状态。

6.2 如何使用 Ansible Tower

6.2.1 安装方法

如果你只是想体验或学习下 Ansible Tower，那么它的安装方法极其简单，只需运行一个 bash 脚本即可。当然，在这个 bash 中，是调用 Ansible 来运行 Playbook 才能做到这么简单的。

all-in-one 版就是所有的 Ansible Tower 的组件，包括服务器、数据库等都安装在一台机器上，一般作为初学者学习和体验的安装版本。all-in-one 版的安装非常简单，只需遵循以下步骤就即可。

准备一台新的虚拟机，安装 Red Hat Linux7 并注册 RHN 账户，或者直接使用 CentOS 7。

1. 下载并安装

◎ 下载安装包。

下载网站的地址是：

https://releases.ansible.com/ansible-tower/setup-bundle/

这里下载的是当前的最新版 3.0.3，对应的安装文件是 ansible-tower-setup-bundle-3.0.3-1.el7.tar.gz。

◎ 复制到准备安装 Tower 的机器中。
◎ 解压。

```
tar xvf ansible-tower-setup-bundle-3.0.3-1.el7.tar.gz
```

◎ 编辑 inventory。
◎ 解压后 ansible-tower-setup-bundle-3.0.3-1 的文件夹中默认的 inventory 如下所示，需要修改 admin_password、redis_password、pg_password 的值为你想设置的密码。

```
[primary]
localhost ansible_connection=local

[secondary]

[database]

[all:vars]
admin_password='password' redis_password='password'
pg_host='' pg_port=''
pg_database='awx'
pg_username='awx'
pg_password='password'
```

◎ 关闭 SELinux。如果不关闭 SELinux，则安装完 Ansible Tower 时可能会报 500 错误。

```
setforce 0
```

为了保证重启机器后，SElinux 仍然是关闭的，我们需要在 /etc/selinux/config 中把 SELINUX 修改为：

```
SELINUX=disable
```

- 运行 ./setup.sh。
- 请注意 Tower 对硬件的要求，如果没有达到下面的要求，则执行安装脚本的过程会中断，并提示硬件没有满足要求。
  - 2GB+内存（推荐使用 4GB 的内存）。
  - 20GB 专用硬盘。
  - 64 位操作系统。
- 如果在安装过程中脚本报错，那么就需要参考更详细的硬件要求，具体请参考官方文档的硬件要求。

  http://docs.ansible.com/ansible-tower/3.0.3/html/installandreference/requirements_refguide.html#ir-requirements

2. 初始化网站

- 修改 admin 的密码。

```
tower-manage changepassword admin
```

- 安装成功，访问你的 Ansible Tower 吧。
- 如果安装的 Ansible Tower 的 hostname 是 ansibletower.example.com，那么访问 https://ansibletower.example.com/。
- 用 admin（用户名）和你修改后的密码登录。
- 第一次登录后，会提示你提供 Licences 文件。
- 申请 Licences。

Ansible Tower 提供了免费使用的 Licences，永不过期，但限制了管理 10 个以下主机，而且不能用一些 LDAP 等高级特性。不过这并不影响我们体验和学习。去这里 https://www.ansible.com/license 申请一个吧。

3. 访问你的页面

第一个页面如图 6.4 所示。

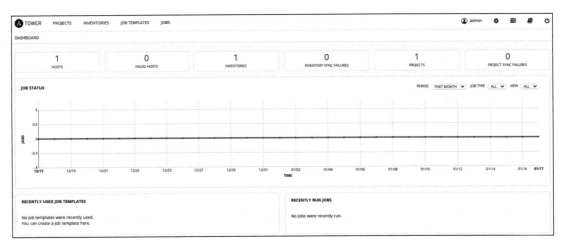

图 6.4　第一个页面

6.2.2　使用方法

在本节，请跟随笔者进入创建第一个 Ansible Tower 的 Job 的旅程吧。

1．预备知识

前面介绍过，Tower 的 Job 是 Tower 中 Job Template 的一次执行，对应 Ansible 中 Playbook 的一次执行。我们先来屡屡思路，然后再来介绍创建第一个 Job 的具体过程。

在 Ansible 中，第一次执行 Playbook，我们需要做什么准备工作呢？

❶ 首先配置连接远程主机的 SSH 连接：

```
ssh-genkey && ssh-copy-id
```

❷ 加入要管理的远程主机，即编辑 inventory。

❸ 编写或者下载 Playbook 脚本。

❹ 执行命令 ansible-playbook helloword.yml。

在 Tower 中创建第一个 Job 与之类似。

❶ 配置连接远程主机的 SSH 连接，即：

◎ 在 Tower 的 SETTINGS 页面创建一个 Credential，Credential 一般包含了通过 SSH 连接远程主机的密钥。

❷ 编写或者下载 Playbook 脚本。

Tower 中的项目对应 Playbook 的文件夹，或者 git/svn repo。创建一个新的项目，包含执行 Playbook 的文件或者 git/svn repo 地址。

❸ 把需要执行一次 Playbook 的信息组合起来。

创建 Job Template，指定 inventory、项目以及选择项目下哪个 Playbook。

❹ 单击创建 Job Template 的执行按钮，就会自动跳转到 JOBS 页面的下面，从而看到实时的执行结果。

2. 理解 Tower 中建立基于密钥的 SSH 连接

用命令行工具创建基于密钥的 SSH 连接很容易，只需执行下面的命令：

```
ssh-copy-id user@remoteserver
```

而用 Tower 创建 Credential，则需要对基于 Key 的 SSH 连接有比较清楚的认识。

如图 6.5 所示，基于密钥的 SSH 连接，需要一对 Key：私钥（Private Key）和公钥（Public Key）。当把公钥复制到远程主机上时，用 SSH 协议并且携带私钥，就可以无须输入密码直接创建 SSH 连接了。Linux 上的 SSH 命令会自动携带~/.ssh/id_rsa 作为私钥，也可通过参数-i 来显式指定用哪个文件作为私钥。

Tower 的 Credential 基于安全的考虑不会直接放在 Tower 所在的服务器上，即使有权限登录 Tower 主机的管理员也没有办法通过任何一个命令看到 Credential 的值。Credential 经过加密后存储在数据库中。在使用的时候再解密使用。

在图 6.5 中：

◎ （1）、（2）、（3）、（4）是创建 Credential 时的步骤。

◎ （a）和（b）是使用 Credential 创建 SSH 连接的步骤。

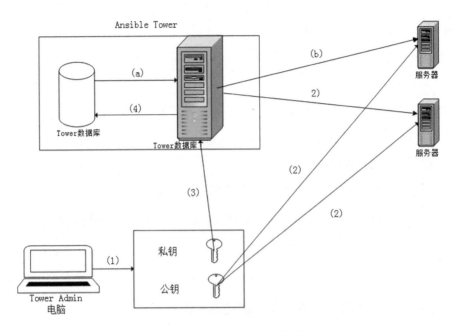

图 6.5 基于密钥的 SSH 连接

3. 创建第一个 Job 的具体过程

我们配合截图来说明创建第一个 Job 的过程，假设有两台虚拟机：10.66.208.194 和 10.66.208.212。

第一步：生成 SSH 的密钥对并创建 Credential。

生成 SSH 秘钥对（Key Pair）——tower_rsa 和 tower_rsa.pub，命令如下。

```
ssh-keygen -f tower_rsa
```

将公钥复制到两个远程主机。

```
ssh-copy-id -i tower_rsa root@10.66.208.194
ssh-copy-id -i tower_rsa root@10.66.208.212
```

用下面的两个命令测试下 SSH 的公钥是否复制成功。

```
ssh -i tower_rsa root@10.66.208.194
ssh -i tower_rsa root@10.66.208.212
```

接下来用截图来展现在 Tower 中创建 Credential 的过程。

❶ CREDENTIALS 的配置在 SETTINGS 页面下。单击右上角的齿轮按钮，再选择 CREDENTIALS。

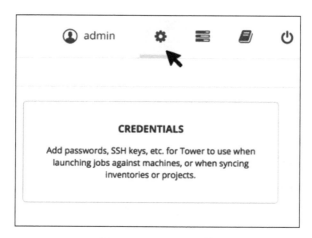

图 6.6　SETTING 页面

❷ 在 CREDENTIALS 页面，单击 "+ADD" 按钮来添加一个 Credential。

❸ 给 Credential 起名字，并选择 TYPE 为 machine。

需要用 SSH 连接的，在 Credential 的类型中都选择 MACHINE；本机连接也选择 MACHINE，不填写 Password 或者 Private Key。更多的 Credential 的类型请参考官网的 Credential Type 定义：

http://docs.ansible.com/ansible-tower/latest/html/userguide/credentials.html#credential-types

❹ 将私钥 tower_rsa 的内容复制到如图 6.7 所示的 Private Key 中，保存该 Credential 即可。

图 6.7　将私钥内容复制到 Private Key 中

第二步：创建 inventory。

这里需要为 inventory 添加两个主机记录，如图 6.8 所示。

图 6.8　添加两个主机记录

❶ 单击菜单栏里的 INVENTORIES。

❷ 在 INVENTORIES 页面，单击"+ADD"按钮来添加一个新的 inventory。

❸ 给新的 inventory 起名字后，单击"SAVE"按钮保存，如图 6.9 所示。

图 6.9　给新的 inventory 命名

❹ 保存后会跳转到下一页，如图 6.10 所示，单击"+ADD HOST"按钮来添加新的主机。

图 6.10　单击添加主机按钮"+ADD HOST"

❺ 添加主机的域名或者 IP，如果是非 SSH 类型的连接则还需要指定额外的连接变量。

Ansible 默认的连接方式是 SSH，如果用 SSH 则不需要添加额外的参数。若连接本机则是不遵循 SSH 协议的，就需要添加一个 ansible_connection 变量来告诉 Ansible 不要用 SSH 协议，而是直接用本地（Local）的方式直接控制本机。

更多的非 SSH 的连接方式可以参考官方文档中的 ansible_connection 参数说明：

http://docs.ansible.com/ansible/intro_inventory.html#non-ssh-connection-types

如图 6.11 所示。

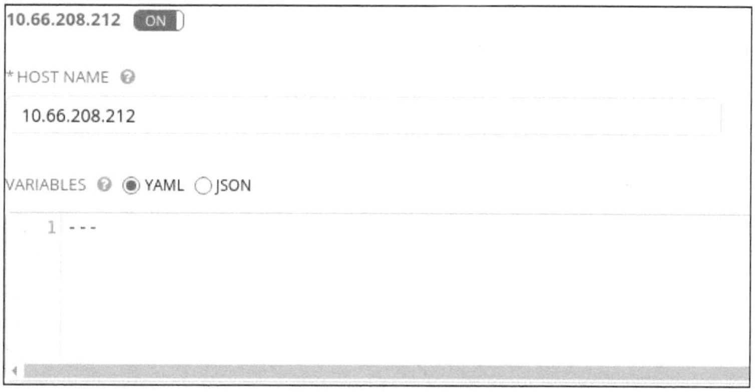

图 6.11 非 SSH 的连接类型

用同样的方法添加另外一台主机。至此，需要管理本机的 inventory 已经创建好了。

第三步：创建 Project。

Playbook 保存在 Git 或者 svn 这样的 SCM 的 Repo 上，这样无论是历史记录的保存，还是共享都很方便，所以一般也都放在 SCM 的 Repository 中。

如果要将 Playbook 放在私有的 SCM 中，则还需要为 SCM 配置 Credential，为了简化步骤，我们将 Playbook 放置在 Github 上，无须特别的权限就可以访问 Playbook 代码。这样在 Project 中，只需配置 Git 的地址就可以了。

❶ 单击图 6.12 中左上角菜单栏中的"PROJECTS"。

图 6.12 单击菜单栏中的"PROJECTS"

❷ 在 PROJECTS 网页中，单击"+ADD"按钮来添加一个 Project。

❸ 命名新的 Project，选择 SCM TYPE 和 SOURCE DETAILS，如图 6.13 所示。

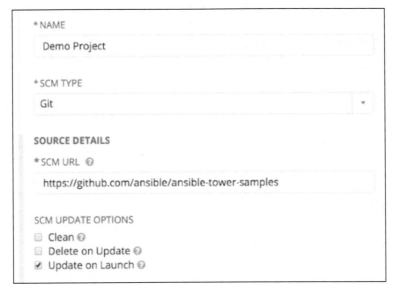

图 6.13　命名 Project

❹ 如果希望 Tower 在每次执行前都去 Git 上更新到最新的代码，请勾选"Update on Launch"，然后单击保存按钮"SAVE"。

第四步：创建 Job Template。

现在我们把所有的信息整合起来，创建一个 Job Template 来告诉 Tower 如何运行 Playbook，如图 6.14 所示。

图 6.14　创建 Job Template

❶ 单击右上角菜单栏中的"JOB TEMPLATES"。

❷ 在 JOB TEMPLATES 页面上，单加"+ADD"按钮添加一个 Job Template。

❸ 命名新的 Job Template。

❹ 选择 PROJECT 和 CREDENTIAL。

单击图 6.15 所示的"放大镜"图标，这样所有的 PROJECT 都会列出来，选择你想执行的 PROJECT，用同样的方法选择 CREDENTIAL。

图 6.15　单击"放大镜"图标

❺ 选择 JOB TYPE，选择运行的 PLAYBOOK 文件，然后单击保存按钮"SAVE"。

图 6.16　选择 JOB TYPE

至此 Job Template 就创建好了。

第五步，执行并在 JOBS 页面查看结果。

❶ 保存 Job Template 后，向下滚动，找到刚创建的 Job Template，单击"小飞机"图标来执行，会自动跳转到 JOBS 页面。

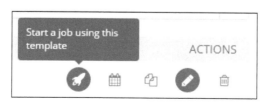

图 6.17　单击"小飞机"图标

❷ 在 JOBS 页面下面就可以实时看到结果啦。

图 6.18　JOBS 页面

❸ 等待 Tower 自动刷新执行结果。

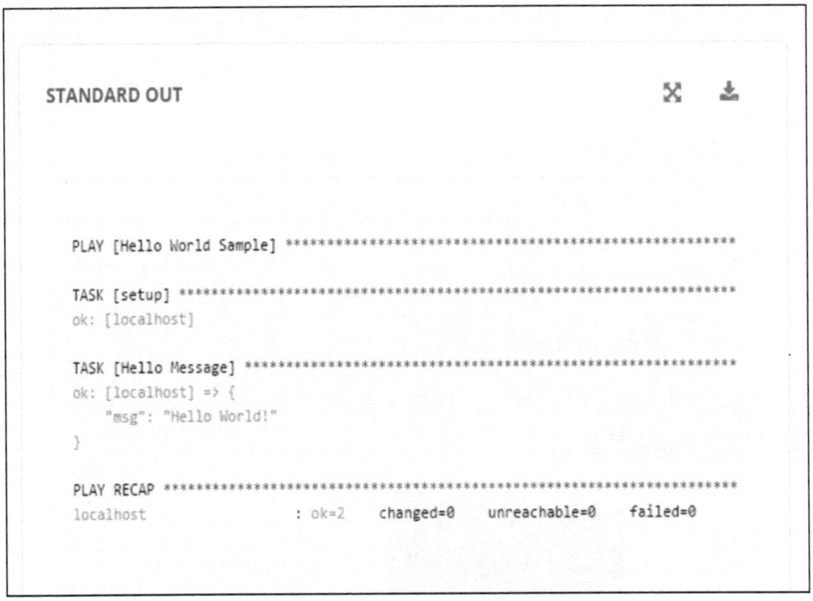

图 6.19　等待 Tower 自动刷新执行结果

4. 更多的小功能：SCHEDULE 和 NOTIFICATION

为了方便用户的使用，Tower 还提供了很多友好的小功能，例如，SCHEDULE 定时任务、NOTIFICATION 发通知给你的聊天软件等。像其他图形化软件一样，多用鼠标熟悉图形化的界面和设计风格，有助于你发现更多的功能。

5. Tower 与 Galaxy

如何在 Ansible Tower 中使用 Ansible Galaxy 中的 role 呢？

如果你有文件<project-top-level-directory>/roles/requirements.yml，那么 Tower 会自动下载 role 到<project-top-level-directory>/roles 目录下。这是为什么呢？因为 Tower 一旦检测到 requirements.yml，就会执行下面的命令：

```
$ ansible-galaxy install -r requirements.yml -p ./roles/ --force
```

6.2.3 总结

1. 谈一谈 Tower 的权限管理

❶ 无处不在的 PERMISSIONS 子页面

PROJECTS、JOB TEMPLATES、INVENTORYS 都有可以设置权限的子页面，Tower 管理员可以根据需要设置权限。

图 6.20　可以设置权限的子页面

❷ 基于用户组的权限管理

为了简化权限管理操作，Tower 在 SETTINGS 页面提供了用户组（TEAMS）功能。

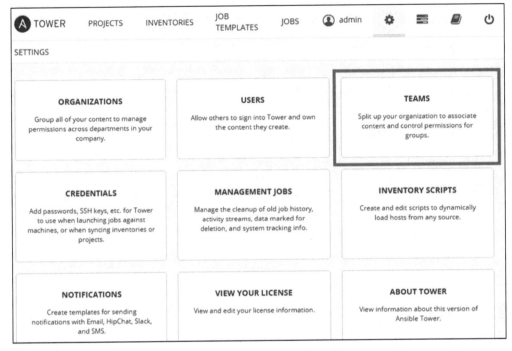

图 6.21　用户组的功能

在 PERMISSIONS 权限管理界面如果赋予用户组某一权限，则该用户组下面的每一个组成员都有相同的权限。

图 6.22　为用户组赋予权限

❸ 权限管理的架构

Tower 的权限管理的结构图如图 6.23 所示，所有的权限管理都是在一个 Organization 下面的。不同的 Organization 之间的权限管理是隔离的，不过因为创建的新 Organization 只能在付费版本的 Tower 中使用，所以在免费体验版中，体验不了多个 Organization 的权限设置。

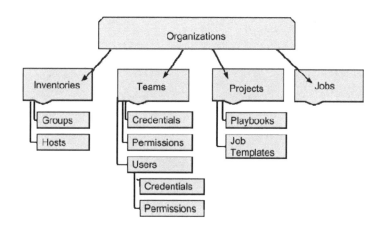

图 6.23　Tower 的权限管理结构图

2. 谈一谈 Ansible 命令行与 Ansible Tower 的关系

Ansible Tower 关于 Ansible 的功能是通过调用 Ansible 命令行实现的，如果登录 Ansible Tower，修改 Ansible 的配置文件，那么对 Ansible Tower 也是生效的。

Ansible Tower 3.0.3 默认会安装 Ansible 2.1 命令行工具。但在该版本的 Tower 中正确配置 Extra 模块后，执行时会一直提示找不到相关模块的错误。这是因为 Ansible 2.1 对 Extra 模块的支持有 bug，登录到安装有 Ansible Tower 的服务器，手动升级 Ansible 到 2.2 以上的版本即可。

6.3　与第三方平台的整合

6.3.1　Ansible Tower API

Ansible Tower 提供了标准的 REST API 来供第三方平台调用 Ansible Tower 的完整功能。为了减少学习 API 的成本，Ansible 提供了一个浏览器页面，供用户查看 API 的用法和返回值。安装好 Ansible Tower 后，假设安装 Tower 服务器的域名为<TowerEndpoint>，那么直接访问下面的网站：

https://<TowerEndpoint>/api

图 6.24　访问网站

6.3.2　Ansible Tower 提供的命令行工具

Ansible Tower 相关的命令行工具一共有三个：

- ansible-tower-service。
- tower-manage。
- tower-cli。

这里不会介绍具体的用法，只是介绍一下使用这些命令的位置和它们的主要功能，如图 6.25 所示。

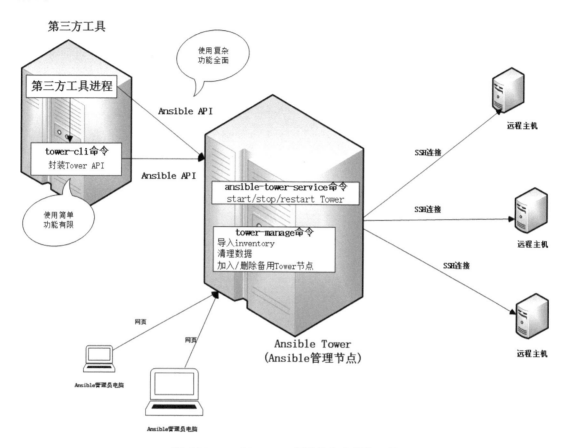

图 6.25　Ansible Tower 相关的命令行和工具

参考资料

所有资源的列表请访问网站：

http://getansible.com/resources

参考的链接列表：

https://en.wikipedia.org/wiki/YAML

http://www.yamllint.com/